The Many Lives
of James Lovelock

Also by Jonathan Watts

*When a Billion Chinese Jump:
How China Will Save the World – or Destroy It*

Jonathan Watts

The Many Lives of James Lovelock

Science, Secrets, Spycraft and Gaia Theory

GREYSTONE BOOKS
Vancouver/Berkeley/London

Copyright © Jonathan Watts, 2024
First published in North America by Greystone Books in 2025
Originally published in the United Kingdom in 2024 by Canongate Books

25 26 27 28 29   5 4 3 2 1

The publisher expressly prohibits the use of *The Many Lives of James Lovelock* in connection with the development of any software program, including, without limitation, training a machine-learning or generative artificial intelligence (AI) system.

All rights reserved, including those for text and data mining, AI training, and similar technologies. No part of this book may be reproduced, stored in a retrieval system, or transmitted, in any form or by any means, without the prior written consent of the publisher or a license from The Canadian Copyright Licensing Agency (Access Copyright). For a copyright license, visit accesscopyright.ca or call toll free to 1-800-893-5777.

Greystone Books Ltd.
greystonebooks.com

Cataloguing data available from Library and Archives Canada
ISBN 978-1-77840-248-7 (cloth)
ISBN 978-1-77840-249-4 (epub)

"Deep blue sea baby" by Gary Snyder, from *Axe Handles*. Copyright © 1983 by Gary Snyder. Reprinted with the permission of The Permissions Company, LLC on behalf of Counterpoint Press, counterpointpress.com.

Jacket design by DSGN Dept.
Jacket photograph by Donald Uhrbrock/Getty Images
Printed and bound in Canada on FSC® certified paper at Friesens. The FSC® label means that materials used for the product have been responsibly sourced.

Greystone Books thanks the Canada Council for the Arts, the British Columbia Arts Council, the Province of British Columbia through the Book Publishing Tax Credit, and the Government of Canada for supporting our publishing activities.

Greystone Books gratefully acknowledges the xʷməθkʷəy̓əm (Musqueam), Sḵwx̱wú7mesh (Squamish), and səlilwətaɬ (Tsleil-Waututh) peoples on whose land our Vancouver head office is located.

For Ani and the NHS

Contents

James	1
Nellie	11
Thomas	31
Helen	46
Christine, Jane, Andrew, John	62
Dian	80
Victor	106
Lynn	149
Barry	187
Sandra	213
Nigel and Bruno	235
Gaia	261
Notes	273
Bibliography	283
Acknowledgements	295
Index	299

James

'He has two loves: the natural world and bombs.'

Love rarely gets the credit it deserves for the advancement of science. Nor, for that matter, does hatred, greed, envy or any other emotion. Instead, this realm of knowledge tends to be idealised as something cold, hard, rational, neutral and objective, dictated by data rather than feelings, even if that requires all traces of humanity to be sucked out of every equation as thoroughly as the air from a vacuum chamber.

The life and work of James Ephraim Lovelock is proof that this is neither possible nor desirable. In his work, he enabled us to understand that humans can never completely divorce ourselves from any living subject because we are interconnected and interdependent, all part of the same Earth system, which he called Gaia. In his personal life, he realised feelings – such as desire, intuition and loyalty – shaped what he was looking for in his research.

He rediscovered relationships – between life and the atmosphere, between humans and nature – that had been lost, or at least neglected, three centuries earlier during the Enlightenment, when ties between human civilisation and the rest of the animate world started to rupture. Back then, a duality emerged of self and other that provided the philosophical basis for the Industrial Revolution by turning nature into a resource to exploit, rather than something that humanity was part of and reliant upon.

Lovelock reconnected us. He put us back inside. He reminded us that we are living agents rather than mechanistic objects. As the former director of the Science Museum Chris Rapley put it: 'Lovelock put the soul back into science.'[1]

This has never been more important than now as humanity grapples with a polycrisis of climate breakdown, extinction of nature and social inequality. Lovelock's theory of interconnectedness takes us to the heart of the world's current problems: our relationship with the Earth – and, crucially, how far we can push it. At its simplest, Gaia is about restoring an emotional connection with a living planet. As Lovelock's closest friend and Gaia theory co-developer, the US biologist Lynn Margulis, put it: Lovelock 'let people believe that Earth is an organism, because if they think it is just a pile of rocks they kick it, ignore it, and mistreat it. If they think Earth is an organism, they'll tend to treat it with respect.'[2]

The author Christopher Hitchens argued no biography was worthwhile unless the subject roamed the high alps of intellectual or ethical advancement. Lovelock wandered at this altitude for decades, scaling the crags of evolutionary theory and atmospheric physics and collaborating or doing battle with many of the greatest minds of the twentieth century: Margulis, Richard Dawkins, William Hamilton, Paul Crutzen, William Golding, Paul Ehrlich, Steve Schneider, Carl Sagan, Christopher Andrewes and Archer Martin. Behind the scenes, he helped to shape the thinking of an oil baron, a prime minister and multiple spymasters, as well as, more prominently, a generation of Earth scientists and environmentalists. As both an apex industrialist and proto-ecologist, he symbolised a wider human struggle to pivot from an abusive relationship with the Earth to one based more on mutual respect and understanding.

This raised the vexed and very modern question of how to circumscribe human activity. Despite Lovelock's personal reluctance to address the question of planetary limits head on, his inventions, discoveries and theories proved this was a necessity if the Earth system was to remain stable. In many ways, this was a more important, if less romantic, contribution to human thought than the Gaia theory itself. The French philosopher Bruno Latour compared the impact of Lovelock to that of Galileo, though he noted they came from opposite directions: the seventeenth-century Italian astronomer and his telescope opened our understanding to an infinite universe, while the twentieth-century English geophysicist and his electron capture detector (ECD) forced us to look inwards at the fragile boundaries of our home planet. British moral philosopher Mary Midgley saw Gaia as an antidote to the atomistic, individualist pursuit of science and an opportunity for humanity to restore a holistic and ethical relationship with nature. The atmospheric scientist Andrew Watson saw his mentor's influence in cybernetic terms: 'Jim was the first person in post-war science to see that the Earth is a system and that to understand the climate of the earth and its history, and the relationship of life and the inanimate earth you have to do systemic modelling.'

While the most prominent academics of the modern age made their names by delving ever deeper into narrow specialisms, Lovelock dismissed this as 'knowing more and more about less and less' and worked instead on his own all-encompassing, and thus deeply unfashionable, theory of planetary life. He took a holistic view. A true polymath, he mastered an extraordinary range of disciplines – chemistry, electronics, medicine, cryobiology, exobiology, atmospheric chemistry, geology, climate science and marine biology – but refused to be hemmed in by any one of them. He even hand-crafted his own instruments on a watchmaker's lathe.

He was a devoted believer in instinct and inspiration, a relentless critic of dogma and orthodoxy and a gleeful breaker of boundaries between academic disciplines, heretically blurring even the distinction between religion and science. Rather than a scientist, he preferred to describe himself as an inventor, which gave him the liberty to play with literary and spiritual allusions in his writings – much to the fury of scientific purists.

Who else would have dared to name their most ambitious theory after a Greek goddess who was more powerful than the Titans? The Gaia hypothesis took Lovelock's thinking far outside the confines of academia and into the realms of environmental politics, popular culture and New Age philosophy. Lovelock would pay a heavy price for invoking this spirit of Mother Earth. The great metaphor of our home planet as a living organism put him on a collision course with distinguished biologists who felt he was leaning too far towards creationism and mysticism. Ford Doolittle believed the Gaia hypothesis was dangerous.[3] John Maynard Smith called it 'an evil religion'.[4] Robert May labelled Lovelock a 'holy fool'.[5]

Most of these early critics later came round to a more appreciative opinion of the theory, though its name — a 'convenient four-letter word', according to Lovelock — remained offensive in staid academic circles, and came to be replaced by the blander, but more scientifically acceptable label, Earth System Science. That is now one of humanity's most important bodies of knowledge, providing insights into the communion of nature and climate, and warnings about the sharply deteriorating condition of both. Numerous institutes across the world are devoted to this study and many of their leading scientists acknowledge they were inspired by Lovelock and Gaia, though this is a love that cannot publish its name. The few academics who are still brave enough to submit research papers using the name Gaia often find themselves excluded from mainstream journals.

Lovelock's great gift to humanity was the repopulation of the atmosphere with life and meaning. He proved the air around and between us is far more than a chemical happenstance or a void to be filled with any old waste we need to discharge. It is, in fact, replete with vitality, the child of billions of trillions of quadrillions of interactions every second. We — humans, animals, insects, plants, bacteria and all other living beings on Earth, along with the weathered rocks of the landscape — are constantly formulating and reformulating the composition of the gaseous outer layer of our planet in a way that keeps it uniquely habitable. That is part of life. And all life dwells within it.

Lovelock's understanding was guided by the invention of an atmospheric monitoring device so sensitive that it could detect trace elements at the level of one in a trillion, equivalent to isolating a single second in a continuum of 30,000 years. With this exquisitely precise instrument – the ECD – the invisible became visible, enabling Lovelock to read the sky, oceans and earth more accurately than had ever been possible, demonstrating their ever-changing chemical complexity and remarkable historical stability. It helped to reveal the steady build-up across the world of petrochemical pollutants, chlorofluorocarbons (CFCs) and other man-made chemicals that posed a threat to life and the stability of the climate. He also used this device to track Cold War enemies, locate hidden IRA explosives and track fallout paths from potential nuclear attacks.

This book applies Gaia theory to a biography by asking whether any idea can be understood as belonging to a single individual. The chapters are organised by the layers of relationships that made up Lovelock's life and shaped his thinking. In this sense, this book is both a biography and an anti-biography. On one hand, it tells the story of a brilliant individual who played a central role in some of the great scientific developments of the twentieth century – the advance of medicine, the search for life on Mars, the discovery of the ozone hole, and the recognition that petrochemicals are disrupting the climate. Yet, at the same time, it challenges the idea of a solitary genius and stresses the importance of interactions – neglectful, manipulative, passionate, loyal, inspiring and resentful. Some were long-lasting and profound. Others fleeting and imaginary. But they all helped to shape the man and his ideas. Those relationships are the essence of this story. Genius does not emerge in isolation. Nor, for that matter, did the Gaia theory.

This helps to explain some of the apparent contradictions in Lovelock's life: why this ecological guru often jumped to the defence of industry, how a self-styled 'independent scientist' was embedded in the British establishment, why his views about the climate flipped back and forth, and why this leading theorist of an organic perception of the Earth advocated geo-engineering and nuclear power.

The chapters on the origins and development of Gaia theory will probably surprise many of Lovelock's followers, as it surprised me. This does not detract from the robustness of the science, but it does challenge – or expand on, at least – the origin myths of the hypothesis. The women in his life, particularly the NASA consultant Dian Hitchcock, played a far more important conceptual role than has been acknowledged until now; so did the Anglo-American military-industrial complex which provided an early cradle for his brainchild. Knowing he did not have long to live, the centenarian told me: 'I can tell you things now that I could not say before.' The true nature of the relationships that made the man and the hypothesis were hidden or downplayed for decades. Some were military or industrial secrets. Others were too painful to share with the public, his own family and, sometimes, even himself. Even in his darkest moments, Lovelock tended not to dwell on the causes of his unhappiness. He preferred to move on. Everything was a problem to be solved.

In person, he was utterly engrossing and kind. We first met in the summer of 2020, during a break between pandemic lockdowns, when he was 101 years old. As global environment editor for the *Guardian*, I had long wanted to interview the thinker who somehow managed to be both the inspiration for the green movement and one of its fiercest critics. His ideas were still extremely influential.

Lovelock and his second wife, Sandy, lived in a former coastguard's cottage within earshot of the English Channel waves crashing on the shingle of Chesil Beach in Dorset. I arrived thinking I should not stay long as he was a very old man who would probably be quickly tired, but Jim astounded me with his stamina, perspicacity and humour. We talked for almost three hours. The encounter was thrilling. Of course, I already knew his work. But among the jewels in his mental treasure trove were anecdotes and insights that I had never heard before. He seemed a living witness and instigator of many of the key events in the history of twentieth-century science and the environmental movement.

Maybe, I thought, he could retrace the knowledge trail that humanity had taken and provide clues as to where we might have made a wrong turn. I asked if I could write his biography. He promised to consider. When I received an answer a month later, it was almost a form of charity. I was in hospital after a cardiac arrest and three defibrillator shocks that had brought me back to life. Lovelock wrote to tell me I should not give up hope because he too had suffered his first heart problems in his fifties yet had gone on to live another half century. And, by the way, he would be very glad if I would consider writing the story of his life.

This became the priority of my Life 2.0. It was a mission and a form of recuperation. On the first of many trips to his home, my wife, Eliane Brum, and I spent a month in a caravan in Swyre, a few miles from Chesil Beach. Every morning I would walk for an hour past the blackberry bush-hedged coast road to West Bexington, then stroll along Labour-in-Vain Lane and down through fields of sheep to Matthew Cottage, where Jim and Sandy would greet me with a warm welcome before we got to work on the task of downloading his life into my notebook. Two, three and sometimes four hours later, I would return, my head buzzing with incredible stories and insights. It is normally a sin for a biographer to claim to know how a subject felt at key moments, but in this case he told me himself, albeit through the prism of later life.

I knew from his last book, *Novacene*, that Jim believed the next stage of evolution was to convert energy and ideas into data.

It seemed I was now part of that process, recording his every word for posterity. 'I know I will die before this is published,' he told me. Rather than a source of gloom, this awareness seemed to be liberating for him. Our relationship became close. Over two years, dozens of visits and more than eighty hours of interviews, I came to see Jim as a friend and was determined to write with an underlying affection and empathy. At the same time, it became increasingly clear from other interviews – of close family, old friends, former lovers, MI6 colleagues, scientific collaborators and current politicians – that his version of events needed careful corroboration. At times, this threw up contradictions, evasions and myths.

I spent many days trawling through the tens of thousands of pages of letters, notebooks, speeches, doodles, invention designs, journal submissions, manuscripts, newspaper clippings, travel itineraries, graphs, photos and confidential documents in the Lovelock Archive of the Science Museum archives, stored at a former RAF base in Swindon. This led me to freedom of information applications to the FBI, and archive requests to NASA, the Jet Propulsion Laboratory, the Margaret Thatcher Foundation and the estate of the author William Golding. Many of their documents have never been made public before. Lovelock also shared intimate love letters, and secret correspondence written in green ink by 'C', the code name for 'Control', the chief of the British intelligence services. Jim's family and friends shared diaries, letters, recollections and descriptions of Lovelock, my favourite of which was by his daughter Jane who said her father 'had two loves: the natural world and bombs'. I thank them all for their cooperation and encouragement to write an unvarnished, warts-and-all biography. To Lovelock's credit, he never challenged this, though he concealed some matters, dissembled on others, and delighted in spinning yarns. I had to be wary because he was a sometimes unreliable narrator and varnisher of his own life.

This was not a problem of mental acuity. Before allowing me to start on this biography, Lovelock was considerate enough to arrange, at his own behest, a dementia test with a neurologist to ensure I would not be wasting my time. It proved unnecessary. There were ups and downs, but overall his mind was remarkably sharp for a man of his age and, while his short-term memory was erratic, his recollection of events in the distant past was often extraordinary. He could recall the Latin names of plants he had examined half a century earlier, reel off the chemical components of complex compounds, recite limericks from the 1940s and recollect the frequency he used as a child to tune in to broadcasts from Sydney on his homemade radio. Recent events were more difficult. Dates were also a problem, though this, he said, was because numbers had been a challenge for him for all of his life. One of the least known and most remarkable facts to emerge from the research for this book was that Lovelock probably suffered from dyscalculia, a form of numeric dyslexia that affects about 5 per cent of people. During his childhood, the terminology of neurodivergence did not exist so there had been no chance of a diagnosis. In school reports, teachers simply chastised Lovelock as weak or lazy at mathematics.

He also struggled to distinguish between left and right, another common trait of those with dyscalculia. This continued throughout his life and made him worry that if he participated in a military drill, he might face the wrong direction and salute with the wrong arm. It also made him a sometimes terrifying driver when he was old enough to own a car, but it encouraged him to look at the world without sides. He was genetically disinclined to divide life in this way. People with dyscalculia tend to be more holistic and creative in their approach to problem solving. According to the British Dyslexia Association, those with this type of brain are often big-picture thinkers who gravitate towards the arts and engineering.[6] This may be why Lovelock had little truck with the left and right tribalism of politics. He felt far more comfortable trying to understand the connections that make up the whole.

That was not true of his personal relationships. Brilliant as he was in so many fields, this was an area where he often felt ill-equipped for self-reflection. Society discomforted him. Despite an ability to charm and a mischievous sense of humour, he preferred to live away from others. Although he developed the world's most sensitive chemical detectors, he could be incredibly tin-eared and inconsiderate of the feelings of those closest to him. Lovelock acknowledged an occasional lack of empathy which he attributed to the loneliness of his early childhood. And for this and much else, he blamed his mother, Nellie. It may just be a coincidence that the man behind the ultimate theory of Mother Earth had an exceptionally troubled relationship with his own mother, but from his earliest years, Jim felt a maternal–filial bond was missing in his life. This was the origin of Lovelock's quest to find and understand the relationships that constitute life, a search for knowledge that was inextricable from a search for love.

Nellie

'I have nothing to do except sit here and think what a mess my poor son has made of his life.'

The year is 1965 and the Gaia theory is floating in the intellectual ether, embryonic and nameless, a brainchild whose parents are only dimly unaware they have conceived. Sure, they feel excited. They know they have hit on a big, genuinely original idea. But they cannot yet comprehend how it will grow and evolve to reshape humanity's understanding of life, to challenge Darwinist theories, to recognise the eternal marriage of climate and nature, to address the question of planetary limits, to soothe and shock global leaders, to tempt and threaten multinational corporations, and to reignite worship of an all-powerful planetary goddess. All of that is still many years away. The Gaia theory gestates slowly. At this stage, the most remarkable thing about the re-conceptualising of Mother Earth is that it coincides with a rupture in the troubled filial relationship of one of its parents.

The year is 1965 and James Ephraim Lovelock is cutting the apron strings. He has recently moved to a new home in the English county of Wiltshire, where his family can be blessedly far from his mother's prying attentions. And after decades as a dutiful son with a steady civil service job, the forty-six-year-old has set himself up as an independent scientist – a freelancer, who lives on annual consultancy contracts rather than a government sinecure for life. His client list is very short but extraordinarily powerful: the US National Aeronautics and Space Administration (NASA), the oil company Shell, the computer firm Hewlett-Packard, the British secret intelligence service MI6 and the chemical firm W.G. Pye. Apart from the fraught relationship with his mother, the English scientist is having the time of his life.

The year is 1965 and Nellie Anne Lovelock can sense momentous changes are taking place, a presentiment that causes her deep distress. It is not just her world that feels out of joint, but her entire solar system. Why? Because her beloved son, 'her Jim', is leaving her orbit. He won't admit that to her, but the seventy-eight-year-old can see the signs, despite her blindness. Jim has ignored his mother's advice to stick with a safe job. He has relocated his family so she can no longer drop in on them. It makes her suicidally depressed, as she reveals in a pitifully harrowing diary entry of that year.

16 January 1965
I have had a bout of the terrible depression about Jim. It seems to me that when I cannot sleep at night that all I strived for has been wasted and thrown away. All he is content to be now is a technician. I cannot quite believe it. I know he did give the impression as a young man that he had a brain well above average. I will not say more but it was such a struggle both for him and us and apparently his academic achievements are just thrown away. It is a sorrow to me.

Nellie is certain her son is wrecking his prospects of a worthwhile life. She considers his wife socially and intellectually inferior and is mortified that he has recently given up a steady government job to become an independent scientist, a mere 'technician' in her class-conscious understanding. For a woman of her age and background, it is so risky as to be almost treacherous. But the hardest blow of all is a personal one. Jim has

Nellie

moved his family out of London to the distant village of Bowerchalke, near Salisbury, which puts him out of her reach. She finds this unfathomable.

4 February
How much I wish I could get to the bottom of the mystery of why my son decided to take himself and his young family into the backwater.

Nell's diary is a testament to an impressively assiduous but utterly despondent mind. The elderly widow never failed to scrawl an entry each and every day, always filling up a half-page, even though her eyesight was so poor she had to read novels in Braille and her mood was often so low that she interrupted her writing with bouts of crying. The journal reveals a strong intellect constrained by convention. Almost every entry starts with a report of the weather (it was a particularly dreary year, even by the standards of South London). Most end with a record of Nell's insomniac sleeping patterns and how often she needs a pill to help her escape anxiety, loneliness and self-pity for a few hours. In the few lines that are not about her son or herself, she is broadly scornful of everyone and everything else.

18 July
Looked at television for 1⅓ hours. Did not care much for it.

Her loneliness begets frustration which begets anger. The diary is seasoned with acid comments and peppered with explosions of rage. Few escape her opprobrium. On 7 April, she throws a pail of water out of the window at a calling cat that is keeping her awake. On 6 August, she argues with the greengrocer. Most frequent are spats with her fellow residents at Pocklington Court sheltered housing. Barbs are meted out with an icy Wildean wit, though the overall tone is heartbreakingly lonely and sad.

1 January

Felt very sorry for poor old Florie but she is so prickly one cannot get near her. Just as well I am on my own. I seem to irritate people.

30 June

Called on Ellen Hughes. Sometimes I like her. Sometimes I do not and this evening I did not like her at all.

20 July

Called on Atkins for a few minutes. Was glad to get away. Poor Mrs Atkins is rather a moaner.

9 November

Went to Miss Hanson and stayed over an hour. It is somewhat of a trial. I've an idea she would rather be on her own.

Nellie suffers from each harsh encounter, but she cannot seem to help herself. This is especially true when it comes to her own family. When she meets her sisters, Kit and Flossie, they always begin with the most pacific of intentions, but there are just too many raw nerves for anyone's patience to hold out for more than a couple of days.

10 September

I do so hope I shall be helped while I am with Kit not to be tempted to say what I think. What a lot of misery I might have been saved in my life if I had kept my wretched tongue quiet. A tongue is perhaps the most dangerous weapon in the world.

23 December

I said the wrong thing at supper time and she [Floss] shouted at me. I do so hate being shouted at. It makes me feel ill.

Nellie

Nellie's volatility explains 'the mystery' of why her son moved his family out of her range and kept himself on the move, because it was Jim – more than anyone else in her life – who determined the mood of his mother. It had been thus since the death of her beloved husband eight years earlier. Tom had been the perfect partner for more than four decades. His loss left a void in her life, which she had hoped her son could fill. But the bond that normally forms between a mother and her child was incomplete, fractured. He did what he considered his duty, but she ached for his love.

A kind word from Jim and she soars into ecstasies. A day or two of silence and she plunges into pools of anxiety and self-reproach. At such moments, Nellie suffers bouts of weeping that can last for hours and leave her exhausted and wishing for death. She frets about her son. Hour after hour. Day after day. Page after page.

18 January
Came back and when alone I start getting depressed, mainly about my Jim.

6 March
Woke at 5.30 a.m. Did not get to sleep again. Start thinking of Jim and that puts an end to sleep.

28 March
Jim phoned. After all the stress and anxiety about him, he seemed quite cheerful and said he would come alone and see me soon. Felt quite ill with excitement.

17 June
If I do not go out I have nothing to do except sit here and think what a mess my poor son has made of his life and what a wasted life I have had. I should not have been born. Oh what a misery I have become.

4 July
Felt very worried. Keep on thinking about my Jim. What is the use of him telling me not to fret about him? As much good as telling the wind not to blow.

This continues for the rest of the year.
During the 365 days of 1965, there are close to 200 entries about her son. Every morning, she goes to the front door to see if there is a letter from him on the mat. In the afternoons, she waits for a call. At night, she lies awake worrying about him and bewailing the absence of a loving bond between mother and son.
In fact, Jim *does* write and call and visit. So does his first wife, Helen, and their children. Over the year, Nellie receives at least twelve letters from him, six visits to her home and exchanges twenty-six phone calls with her son and his family. She also visits them twice in Bowerchalke for a total of eight days. But for Nellie, it is never enough. She is sure he is hiding something from her. Eventually she concludes she is incapable of eliciting affection.

10 June
I am wanted by nobody in this world. People are so kind. SO kind. They always want to do the giving. I apparently have nothing to give that is wanted, least of all by myself. The horror of having to live many more years like this. Roll on Death. Have committed the worst sin now. Self Pity.

Nellie's visits to Jim and Helen's newly built home in Bowerchalke are the moments of greatest tension in the year. The build-up is always filled with a mix of anticipation and dread.

30 March
I should be looking forward to going to Jim's but am getting so afraid of the nervous strain it entails I find I cannot stand it now.

Nellie

Once she arrives, Nellie is too wound up to enjoy the moment. During a disastrous visit in April, she accumulates a mental list of sleights, real and perceived, all of which she pours out into her diary: disappointment that Jim was too sick to meet her at the station; disapproval that Jim and Helen escaped to the pub for a 'beer session' during her stay; irritation that talks with Jim were interrupted by other family members; frustration at her son's tendency 'to skirt around' her questions; long gaps between meal times that left her 'starving'; and, worst of all, the upsetting arguments that erupt when the three generations of the family are gathered together.

The misery for all involved seeps through the pages. To make matters worse, on Nell's final day she cracks her head on a counter in the dining room. She leaves with a painful bump on her head and feeling 'weepy and depressed' inside. What really hurts, however, is the distance with her son: 'It seems I shall never have a complete understanding of Jim. He avoids it.'

The situation improves somewhat in the second half of the year, after Jim tries to reassure his mother that his work is going well and that he has had an essay about alien-life detection accepted by a prestigious scientific journal.

15 July

Jim came to breakfast at 8.00 and we talked steadily until 10.00. He is very sweet and thoughtful when he is alone with me and does his very best to reassure and allay my fears and anxieties. It is too difficult to understand all the work he is connected with. Can only hope things turn out as well as he seems to think they will. He has a paper being published in *Nature* in a week or two. It will be in connection with his 'space' business.

Though nobody, including her son, is aware of it, this is a historic moment. That paper in *Nature* – 'A Physical Basis for Life Detection Experiments' – is the publication that will provide the foundation stone of the Gaia theory. It is the birth of one of the century's boldest ideas. But Nellie has other things on her mind.

In the following months, she remains depressed and pines for more communication with her son. As 1965 ebbs away in a flurry of snow and widening darkness, Nellie becomes colder and more bitter. The electric bar heater in her flat stops working. She increasingly argues with the other residents in the care home. Days that should be celebrations pass with regrets.

18 December
My birthday. It rained all night and has rained all day without ceasing. No post for me at all. I am aware of what a strange family we are.

It is the same story at Christmas and New Year. Her son does not call or write. Nellie ends the year alone, knowing something fundamental has been missing in her relationship with her son since his birth.

When she pushed Lovelock out of her womb on 26 July 1919, Nellie suffered twice. First from the physical pain of parturition. Then, a few days later, from remorse as she gave up the only child she would have in her life to her mother. She later told Lovelock she had not wanted a baby but had wept inconsolably for an entire night after handing him over.

She and her husband, Thomas, had made a choice. They had married four years earlier with almost nothing in common except their love for each other, an interest in art and a resolute determination not to be encumbered by children. They both had bitter memories of looking after dependents. In Nellie's case, it was her siblings. She was the eldest of five children in a lower middle-class family in Islington, who had social aspirations but little money.

Her mother, Alice Emily March, was determined to marry off her four daughters as well as possible, but this required the acquisition of clothes, manners and accomplishments that the family could not afford on her father Ephrahim's income as a bookbinder. It was decided that Nellie should be sacrificed to this cause. Although she had shown great academic promise at school and won a scholarship to a blue-stocking grammar school, she was told it was her duty to look after her younger siblings. No doubt, her mother also considered that as well as being the most intelligent, Nellie was also the most argumentative and independent-minded of her daughters and thus harder to mould. At fourteen years of age, she was sent to work in a factory, where she spent most of her teenage years. Her salary was used to ensure her sisters, Florence, Annie and Kate, could be brought up as young ladies.

During the early phase of the First World War, her mother would stage dance parties at Letchworth Hall for officers who were about to be sent to the front. Given the horrendous toll in the trenches, this was close to a death sentence, so passions ran high. With no time for extended engagements and social scrutiny, the proposals came thick and fast. In quick succession, all three of Nellie's sisters married upwards: Kitt into the wealthy Leakey family, Ann to a tobacco company executive and Florrie to the owner of a tailor's shop. The only son in the family, Frank, was left to fend for himself. For their mother, this was a triumph, but for Nellie it widened the status gap with her siblings. While her sisters were living like the heroines of a Jane Austen romance, she felt more like Cinderella. It was grievously, painfully unfair. The disparity generated a resentment that would fester for decades. 'She was cruelly cheated ... She felt very bitter about it,' Lovelock remembered.

The injustice fired her political passions. Having been treated as social cannon fodder, Nellie closely identified with movements that challenged inequality and demanded more rights for the disenfranchised. She considered herself a suffragette and a socialist, devoured the works of George Bernard Shaw, smoked cigarettes as a symbol of her independence, and – in the ultimate act of rebellion against her mother's values – hooked up with a man that she knew her family would consider socially inferior.

Thomas Lovelock was the opposite of what was considered a good match in the social values of that period. He was a married father of two, a former convict, an amateur boxer and had grown up in an illiterate family of agricultural labourers. On 12 June 1915, a year after his first wife died, Tom and Nellie married. Their wedding photo shows a dapper couple: Tom in a straw boater, with gloves and James Joyce glasses; Nell in a dark, heavy dress and a floral hat with a veil. They were in love and wanted to enjoy their liberty together, unencumbered by children. Then disaster struck. Nellie became pregnant.

It took a very special moment in history to make the couple drop their guard. The end of the First World War on 11 November 1918 was an occasion for raucous celebrations. After four years of tension, misery and death, London exploded with relief, joy and life. Some 20 million soldiers and civilians had perished on the battlefields of Europe, and 800,000 of them were British – the nation's highest toll in any war before or since. The Armistice Day cessation of this slaughter released a wave of euphoria. Huge crowds gathered in Trafalgar Square and outside Buckingham Palace, setting off fireworks, hanging off lamp posts, drinking till dawn and releasing all the pent-up tension of the previous four years. Caution went by the wayside. Tom and Nellie were not the only ones to let themselves go. The birth rate in the UK increased more than 20 per cent from 1918 to 1919.

Tom and Nellie's feelings were ambiguous. The war years had been good to them. No one in their close family had been sent to the front. Both had escaped oppressive family ties and improved their lot in life. Nellie was flourishing. No longer carrying the burden of her sisters, she had trained as a clerk and showed such prowess that she was appointed personal assistant to the head of Middlesex County Council. Tom was also starting afresh after the misery of his first marriage. He now had secure employment in the gas works, which was then a modern industry.

Nellie

The couple had reason to believe the future would be better for people like them. The women's rights movement had gained momentum during the war because so many men had been sent to the front. Class dynamics were changing in the wake of the Bolshevik revolution in Russia. And, on a personal level, they had found a kind of liberation in each other. In photographs from this period, they appear happy.

The end of the war meant they could concentrate on their shared dream of opening an art shop in Brixton and taking holidays together in Europe. The opportunities seemed boundless. Then Jim arrived. 'For my mother, pregnancy was a disaster,' Lovelock said ruefully.

Nellie decided she would rather let go of her son than her dreams. She would bear the child, but she did not want to raise him. Her mother would do that. It was, after all, only fair that Alice repaid the sacrifices her daughter had been forced to make for the sake of her sisters. Lovelock was born in his grandparents' home in the leafy suburb of Letchworth Garden City. As soon as Nellie recovered, she left to join her husband in Clapham, leaving her son behind with his grandparents.

For the next six years, he was to grow up in their four-bedroom villa, Norton Croft. His mother and father would visit only at Christmas and a few other rare occasions. The rest of the time he was in the care of the old couple. His grandfather did not want another mouth to feed. He preferred to play cribbage than put up with the youngster. In fact, he was resentful of the attention his wife would lavish on the child. This was captured in an expression the elderly man would repeat whenever he was piqued with envy or frustration: 'Jim first, Jim second, Jim a close-run third.' Jim's cockney grandmother – Nana, as he called her – fed him, bathed him, put him to bed and made him feel loved. But she was too busy to watch out for him all the time.

Instead, he was allowed to wander and wonder alone, which set the habits of a lifetime. Without asking permission, the golden, curly-haired child would amble over to the open woodland near Letchworth's common. Occasionally there would be other boys and they would dare each other to jump over wasp nests, but he did not have regular friends. More often, his only company were plants, insects and birds. He would sometimes get so distracted during his rambles that he forgot to go home for dinner, prompting his grandparents to send out search parties. He did not mind getting a little lost from time to time and thought nothing of dawdling around the town and even begging money from strangers to buy sweets. He considered himself perfectly happy: 'I was alone until I was six. That has advantages as well as hang-ups. It made me curious and that was the basis for my character. I was never afraid.'

That started to change when he was old enough to enter primary school. Lovelock was suddenly immersed in a world of playmates, where his lack of early socialisation became apparent. Other children were already skilled in playground charm and guile, which he did his best to emulate but was never quite comfortable with. This wasn't his only challenge. The teachers realised his mind did not work the same as those of his fellow pupils. He could not tell the difference between left and right – a disability that was to stay with him for the rest of his life. He also exhibited what would today be called behavioural issues. Back then, he was simply described as precocious and spoiled. He was certainly reckless and ended up being expelled for trying to persuade a young girl in another class to eat hemlock and other poisonous plants that the teacher had warned them to avoid. He also nibbled a deadly nightshade himself, which induced vomiting and delirium, the first of many run-ins with the education authorities over health and safety regulations. 'Without doubt I was a spoiled child and sometimes dangerously mischievous,' he admitted.

When his grandfather retired, they all had to find new lodgings. This forced a reunion with his parents. They moved in with Nellie and Tom, who were then living above the art gallery they had opened in Brixton. After the spacious, cosseted life of liberty under his nana's care in leafy Norton Croft, the relocation to 226a Brixton Hill, with its urban pollution and parental supervision, came as a shock.

The three generations shared an expansive two-storey building that housed the gallery, a workshop and quarters for a rent-paying tenant to supplement the family income. Customers would wander in throughout the day and browse works by local artists hung in the high-ceilinged, hessian-carpeted hall. Nellie and Tom slept in ground-floor quarters behind the shop, next to a large gas-lit living room where the family would gather to eat, play cards or listen to popular opera on Nellie's gramophone. Lovelock, his grandparents and the tenant had their bedrooms on the floor above the gallery looking out on the street. Bathtime for Lovelock was in a tin tub in front of the ceramic gas heater in another spacious room behind the gallery. The kitchen and toilet were housed in a brick shed in the backyard. The draughty structures could be bitterly cold in winter. Glasses of water left on the bedside table would be partly frozen in the morning.

His parents were considerate, after a fashion. They looked after Jim, but he felt they were distant. Tom and Nellie preferred to spend their time with each other and their energy on running the shop, Brixton Hill Galleries, which provided framing services and touch-ups, along with sales of prints and paintings. An art gallery was a risky venture, but in the late 1920s, the gamble looked like it might pay off. Brixton was booming in the post-war era and, in the UK, its department stores were second only to those in Oxford Street.

Business at the gallery was brisk. Nellie kept the books, fielded enquiries at the counter and established relationships with artists. During the day, Tom walked his rounds, collecting gas-meter money, then returned in the evening to touch up or frame paintings in the basement workshop. Lovelock remembered his parents 'working like dogs' in those years, but it paid off. They were able to take on extra help – a disabled war veteran named Wetherby. Lovelock sometimes served as an errand boy. He grew to hate the sound of the telephone – then a fashionable new contraption – because the ring was usually followed by an order from his mother to fetch or deliver something.

'She maintained a bad relationship with me and was always sending me out on errands. "Oh Jim, can you run to the butcher?" I didn't mind the errands, but I was desperately upset that she did not want me around. I didn't feel they were making an effort to push love in my direction.'

Once again, Lovelock had to amuse himself. 'I was allowed to go anywhere in London by myself. In fact, my parents were very pleased when I pissed off . . . My childhood was one of an extraordinary degree of loneliness. I wasn't unhappy. Anything was better than school.' On Sundays, the one day of the week when the family could be together, his parents took him on the tram to Kensington, but as soon as they arrived, they went their separate ways. While Nellie and Tom perused the art in the Victoria and Albert Museum, their young son crossed the road and spent a solitary day wandering around the Science Museum. When the doors closed in the evening, he would wait outside for his parents at a teashop and they would go home together, all quite satisfied in their very different ways.

This separation was by mutual agreement. Young Jim naturally gravitated towards chemistry, physics, biology and mechanics. The Science Museum was his playground. He would spend hours exploring every corner, his young mind soaking up knowledge that was beyond most adults. For years afterwards, he would make a pilgrimage to this cavernous institution, scrutinising Newcomen's atmospheric steam engine, a Van de Graaff generator that produced huge sparks and an antique polarimeter that rotated the plane of coloured light frequencies. He recalled an exhibition on mining, which had sticks of dynamite and simulations of explosions: 'As a potential young terrorist I absolutely lapped this up; this was fascinating stuff. If only I could get my hands on this, what fun we'd have at Guy Fawkes Day.' This love of bombs and spirit of playfulness made him determined to become a chemist.

During the week, he was enrolled in Dane School, a private educational institution close to Brixton Prison. His underdeveloped social skills once again made him something of an outsider among the other pupils, while his oddly wired mind frustrated and delighted his teachers by turns. He found most basic mathematics impossible. As well as his constant confusion between left and right, which handicapped his manipulations of the two sides of even the most basic equations, he was incapable of holding two numbers in his mind at once, which meant he struggled to carry digits over into different columns.

Yet, his brain made up in memory what it lacked in processing power. Lovelock had near-eidetic recall, which meant he learned fast and could recite lessons months later with astonishing accuracy. Although he had trouble with basic addition, his multiplication was first class because he could easily recall the times tables up to base thirteen. This – and a tendency to constantly question what he was told – gave him a reputation of flippancy among the staff. 'Once the teachers turn against you, you are lost,' he later said. During the first year, he was caned so often across the knuckles that he briefly stopped going to school in protest.

In his spare time, the young lad sketched scenes from the local neighbourhood and familiarised himself with shops on the street: Wheatcroft and Cumper confectioners, Albert Edward Callaby's furniture dealer, the undertaking service of Maxwell Brothers, Barclays Bank, New Imperial Pawnbrokers and Inger Walter auction house. Then he stretched his horizons. London buses were cheap; a four-penny day pass would take him anywhere in the city. Jim was small even for his age, taking after his 4-foot 10-inch mother, but carried himself with self-assurance and a fearless certainty that he could wander undisturbed.

Lovelock was already showing signs of intellectual precociousness. The young boy picked up chess so quickly that his father used to invite drinkers at the nearby Telegraph pub to challenge his son to a game and bet pints on the outcome. Lovelock built his own shortwave radio receiver with a kit contained in a children's hobby annual he received for Christmas around the age of ten. To upgrade it, he sold his stamp collection and used the money to buy capacitors, an electron tube and long lengths of wire that he would wind around a pencil to make frequency chokes. Thanks to this tinkering, he was able to pick up broadcasts of 'The Internationale' from Moscow and sermons from the Vatican. At one point, he was able to reach the Australian VK2ME radio station and called the whole family into the room to listen to the Sydney weather forecast. This early insight into gain and feedback would, many years later, prove an asset when he entered the field of cybernetic analysis.

Nellie took him with her on the half-mile walk to the library. Starting with Jules Verne, Olaf Stapledon and H.G. Wells, he tore through the science fiction collection – the start of a lifelong passion – and then asked if there were more serious books. The astonished librarian peered down at the small boy before directing him to the basement, which held bulky specialist tomes such as Wade's *Organic Chemistry* and *Organic Physics*. Lovelock devoured the tables of data, illustrations of the apparatus and instructions on how to make hundreds of different compounds. A particular favourite was the scientist's description of the preparation of methyl iodide, which ended by shaking the glass to reveal glints of 'liquid jewellery'. The young boy was enraptured. 'I took to Wade's *Organic Chemistry* in a very similar way to which a religious child might take the Bible, as something to absorb and gain things from.'

Not all of the material was highbrow. His mother would scold him for buying 'rubbish' pulp-fiction comics from the local newsagents, but he read every spare moment of the day. Between classes, on the bus, in the toilet and, surreptitiously, late at night by candlelight. On one occasion, he was so absorbed in his books that he didn't notice a sheet was too close to the flame and had to extinguish a small fire.

His favourites were the works of his boyhood hero, the geneticist J.B.S. Haldane. Here was a scientist after Lovelock's heart, who never flinched at trying dangerous experiments on himself. Decades later, Lovelock would remember the joy of recognition he felt while reading Haldane's account of swallowing ammonium chloride 'until his liver fizzed' or putting himself in a divers' decompression chamber to investigate the bends until his eardrums burst, leaving a hole in his ear through which he could blow cigarette smoke. This was exactly the kind of risk Lovelock was to take again and again throughout his life in the name of science.

Although Lovelock felt neglected, it is evident that Nellie took care of the basics. She made sure he was well fed and turned out, and she took him to Lyons teahouse on his birthdays for his favourite meal of liver-sausage sandwiches and chocolate éclairs. She far-sightedly registered herself and her son as Quakers to ensure he would never have to fight or kill. During the First World War, she had learned while taking notes at military service appeal tribunals for Middlesex County Council that members of this religion were exempted from fighting. The gatherings at the Brixton Friends House stirred something spiritual in Lovelock, who would never forget the Quaker mantra that God is a 'still, small voice within'.

Determined to move the family up in society, Nellie also gave her child the best education the family could afford. At the age of nine, Lovelock was enrolled in the Strand School, a fee-charging red-brick grammar school for the sons of doctors, lawyers and other middle-class professionals that was set up as an educational conveyor belt into the civil service.

He resented the step up: 'My mother was living, through me, her failure, and here was the chance to go to grammar school.' The headmaster, a Reverend A.C. Digby, used to make the youngest boys sit on his lap, which made Lovelock feel uncomfortable. He was often caned and started wearing three pairs of underpants to soften the blows. Teachers thought him careless and often placed him near the bottom of class rankings, especially in mathematics. At the end of the first year, his school report stressed how difficult the settling-in period had been. 'He made a shriekingly bad start,' wrote his form master, Pitson, in the Lent term school report for 1930, ranking Lovelock twenty-ninth out of thirty students even in his favourite subjects, chemistry and nature study. The history teacher deemed him 'lazy'.

Worse, Strand revealed the trap of his mother's social aspirations. As the son of a shopkeeper, Lovelock did not dare to bring classmates home, in case they found out his family were 'mere trade'. He became so insecure about his working-class accent that his uncle Leakey felt obliged to provide *Pygmalion*-style elocution lessons that pinched his vowels and clipped his consonants in the manner of a well-bred gentleman. On one occasion, a teacher asked the class how their parents voted. Lovelock was the only boy who answered Labour, much to the derision of his classmates. Nellie had helped his academic prospects enormously, but he would always blame his mother for what he later described as the most miserable period in his life.

His poor health made matters worse. As well as horribly frequent bouts of the usual childhood diseases – measles three times, mumps twice and chickenpox four times, by his own tally – Lovelock suffered severe bronchial ailments, which were all too common in 1920s London due to the mephitic shrouds of coal smoke that frequently enveloped the city. Lovelock's contemporary, the historian Eric Hobsbawm, described the London smog of the mid-war era as 'a world of ten metres circumference. Beyond that a whiteness that sucks everything up . . . The fog wells up before me, sucking, shutting everything off, above, everywhere: white sleep. Shadows.'[1] Making matters worse, his father and mother both smoked prodigiously, creating a pall inside the living room that could almost compete with the haze in the street. The local doctor would often cycle to the house to listen to their child's wheezing lungs, but his healing powers were limited. Penicillin was discovered in 1928 at a hospital just half a dozen miles away, but it would not come into common use for another decade. Asthma inhalers would not be invented for another thirty years. Back then, the usual prescription was chloroform-based cough syrup and extended rest. In his first three years at Strand, Lovelock was absent due to sickness for more than three months.

On the doctor's advice, Nellie and Tom packed their son off to the countryside during every school holiday. First to aunts and uncles, then to farmhouses in East Anglia, Essex and Dorset that would offer cheap lodgings for a few weeks. Each stay was a lottery. The best was an open-hearted family who taught him how to ride a bicycle. The worst were stern puritans in Little Downham, a village set amid the flat, drained marshes of Cambridgeshire. Lovelock felt depressed by a landscape that was devoid of contours and colour, and oppressed by his religious hosts who forbade any reading except the Bible and any hike except to church. 'It was repressive and horrible in all sorts of ways. And the countryside seemed to reflect it. You could just look for miles and hardly see a tree anywhere,' he remembered.

The young boy frequently had to cope alone, sleeping in unfamiliar bedrooms and sharing meals with strangers. It was from around this time until puberty that he started to have a recurring dream of walking alone on a white plane with parallel black lines menacingly drawing nearer and nearer as a babble of hostile voices grew louder and louder. This geometric nightmare took such a strong hold on Lovelock that it filled his imagination even during the day. 'It could be scary,' he recalled. 'One moment I would be walking down a country road and the next, I'd be in this dream world, then I'd be walking again. It spooked me.'

He came to dread the holidays. His parents, meanwhile, lived for their annual vacations together on Thomas Cook tours of

European centres of art and music, collecting passport stamps from Rome, Paris, Zurich and Madrid. Lovelock felt health concerns were not the only reason he was shuttled off to distant farmhouses in the home counties while his parents paraded across the continent. Nellie's idealism tended to be social rather than personal. Lovelock recalled the petite figure of his mother courageously berating burly cart drivers for whipping their draught horses too savagely as they struggled up the steep Brixton Hill. She would speak out about women's suffrage and labour rights. In the evenings, she would often go out to popular operas and exhibition openings with a wealthy, much older bachelor named Charlie Wright.

Her husband did not seem to mind. Tom was too laid back to be envious, and after a double workday of coin collecting and then labouring in the basement, he was glad of the time to relax with a copy of the *Racing Post* and a pint at the Telegraph. Nellie aimed higher. Her ideals were progressively left-wing and feminist, but she had conventional ideas about private education and correct enunciation. Brought up in the East End, she masked her natural cockney to try to come across as more refined but ended up speaking with what was known derisively at the time as a 'Kensington accent'. Lovelock was mortified. But she was doing her best to move up in the world, and to take her son with her.

The young Brixton scallywag – or, as Lovelock later described himself, 'the obnoxious little boy' – in the 1920s would not have been able to identify let alone appreciate his mother's efforts, nor could he articulate the effect of her emotional distance from him. Back then, he said later, he felt no yearning for love, nor even a sense of injustice about being left behind. It was only in later life he would consider the consequences: a shortcoming of guile, street smarts and empathy. 'If I hadn't spent so much time alone as a child, I'd be a better person,' he reflected. 'I'd be better at understanding other people's problems.' For this, he blamed his mother. His father, Tom, could do no wrong.

Thomas

'A feeling for the Earth.'

A cold fury seethed in the breast of the teenage Lovelock as he peddled through the undulating, hedgerow-flanked roads of the North Downs in Kent, a homemade bomb wedged under his seat. The year was 1935 and he was on a mission to right what he felt was an unpardonable wrong: the fencing-off of a stretch of the ancient Pilgrim's Way footpath that stretches from Winchester to Canterbury.

The young rebel was furious at the infringement of his right to roam. This was a major concern for many of his contemporaries and a symbol of resentment against the landed aristocracy. Protesters, particularly in the more class-conscious north of England, had recently organised mass trespasses into areas of great natural beauty that had been closed off to the public. In 1932, the imprisonment of six people who had dared to climb Kinder Scout in the Peak District caused a national outcry. Lovelock's father sympathised with their cause and passed on his feelings to his son, who needed little encouragement to try out his chemistry skills.

In his family's kitchen in Orpington, he had already prepared the materials he needed: a yellow sludge of picric acid (trinitrophenol) for the explosive and white crystals of phenol (carbolic acid) for the detonator. Today, both compounds have long since been banned from regular sale because they are toxic and highly combustible, but in 1930s England, they were available over the counter at ironmongers and hardware stores. A teenage bombmaker like Lovelock had no problem securing abundant quantities, which were decanted into corked bottles. He carefully measured out the proportions he had learned from chemical manuals borrowed from the library and then stashed everything he needed in a small bag under the seat of his sturdy BSA bicycle.

Pedalling to his target near Chevening took almost an hour along mostly deserted country roads. When he reached the hill above the stately home, he stopped, checked that nobody was around, unpacked the bag, took it slowly down the slope on foot, packed the bomb at the base of the fence, lit the fuse, ran off and then watched with joy as the offending obstacle was obliterated with a satisfying roar and a billow of smoke that would have been visible from the house.

Having made his point, Lovelock rode home delighted. If the teenager had been caught, he would almost certainly have been charged with a criminal offence. By today's standards, the detonation of such an improvised device might even result in him being branded an eco-terrorist. 'I was really bad then. I could have ended up in a childhood jail,' he reflected. 'I can see why I thought the fence was outrageous, but it was wicked of me. Somebody could have walked by. It was bloody-minded.'

But the gesture was entirely in keeping with his passions of the time and hinted at the vocation that would follow. Lovelock was a pacifist bombmaker, a spiritual Quaker fascinated by the explosive power of chemistry. 'The craft of making a good bomb is all about how to make a good detonator that won't blow your hand off,' he said, before adding with pride: 'In a certain way I was a natural chemist. It's like being a good cook. Not many people made bombs. And, of those that did, not many still have all their fingers.'

It was neither Lovelock's first nor his last explosive. From the moment he read Wade's *Organic Chemistry* at the age of nine, he was fascinated by bombmaking. As a young boy, he could manufacture his own nitroglycerin and on Fireworks Night, 5 November, he would add a little extra oomph to the rockets and Roman candles with items he had picked up at the shops or filched from the school laboratory. His newfound power to make things go up in smoke with a very loud bang was a guaranteed way to impress his peers and, more importantly, his father. Thomas Lovelock was delighted by his son's prowess and egged him on. Lovelock recalled digging a hole in the back garden, burying his biggest bomb yet and watching it explode so forcefully that it shook the entire house and woke the neighbour. His mother was aghast. His father applauded.

The family had moved to Orpington, a leafy London suburb on the border with Kent, after being forced to give up their art shop during the Great Depression. The ripples of the Black Tuesday stock market crash in the United States had taken months to reach Brixton, but they struck with disruptive force. First, the flow of customers reduced to a trickle, then they dried up altogether for more than a year. His parents laid off their one employee, the former veteran Wetherby, but this cost-cutting was not enough to save the business. The family's income had shrunk to Tom's salary as a gas meter collector, which was insufficient to pay an inner city rent for a gallery. It was the end of Nellie's dream to be something in the art world. They sold up, retired and moved to the countryside. This meant Lovelock was now forced to spend more than two hours every day commuting to and from the Strand School, but he was delighted. He no longer had to fear being outed as 'trade', he was closer to the countryside and finally he was able to spend time with his father.

Lovelock adored Tom as much as he resented Nellie. This emotional judgement was unjust, but fairness did not come into it. His mother had nursed him through ill health and ensured he had a good education, but he never forgave her for leaving him alone during his infancy. He saw Nellie's social pretensions as an oppressive force in his life, and came to associate her with artifice, strain and the heavy constraints of civilisation. His father, on the other hand, was beyond reproach as far as he was concerned.

The record might show Tom was a woefully irresponsible man who had committed his first wife to an asylum, abandoned his first two children and then turned his back on the baby Lovelock for the first few years of his life. But his son ignored that. It was easier to enjoy the company of a father who encouraged and praised him, compared to a mother who tutted and sighed. In Lovelock's eyes, Tom was a hero, whose only weakness was an easy-going character that allowed him to be led astray by more forceful personalities, particularly Nellie. Lovelock came to see his laid-back father as the embodiment of nature and the virtue of letting things take their own course.

Born in 1873 on the Berkshire downs near Wantage, Tom was one of thirteen siblings in a poor rural family. In the early 1880s, they lived in Kiln House, the lodging of the brickworks where his father, James, was foreman. He died soon after and, for a while, the family had to survive on little more than turnips donated by a sympathetic farmer – or so, at least, went the family's oral history. Tom, then a teenager, tried to feed his siblings by poaching rabbits from the estate of the local gentry. He was caught, convicted and sentenced to six months in Reading gaol, where he was forced to spend six hours each day silently walking on a treadmill while staring at a wall. The poet Oscar Wilde, who was incarcerated in the same prison a decade later, described the misery of this most Victorian of pointless and exhausting atonement machines.

Thomas

> With midnight always in one's heart,
> And twilight in one's cell,
> We turn the crank, or tear the rope,
> Each in his separate Hell[1]

True to Lovelock's description of his father as 'tough as old boots', Tom emerged unbroken from this and subsequent hardships. He fled to London, one of a throng of rural migrants trying to re-invent themselves in what was then the wealthiest city on Earth. He found lodgings in Clapham, ironed out his farmhand's drawl, landed a job in the South Metropolitan gasworks in Vauxhall and enrolled in Battersea Polytechnic, where he learned to read for the first time.

His first marriage was a calamity. After his wife, Harriet, had two children, she was certified as an imbecile, though by the standards of the age that may just have meant she had a learning disability or was suffering from postnatal depression. How much Tom tried to support her is unknown. He later told Lovelock that he did not consider himself capable of working and raising the children alone, so he put the infants into the care of a charitable children's home. Harriet was committed to a pauper's asylum and, for the next decade, she was shuttled between institutions in Camberwell, Chartham, Colney Hatch and Leavesden, where she died in 1914.

Tom was not one to dwell on remorse. Nothing seemed to bother him. He would escape into nature and the simple pleasures of life. Always fascinated by how the world worked, he encouraged his son's innate curiosity. When Lovelock was four years old, Tom – on a rare family gathering for Christmas – gave him a box of wires and electronic odds and ends to play with. He would take Lovelock to visit his half-brother and sister, though Nellie soon put a stop to that. Tom later taught Lovelock to identify the tracks of badgers and foxes, and the importance of patience if he wanted to see them emerge from their burrows and lairs. He would take great delight in feeding wasps with marmalade on his fingers. Many years later, Lovelock was to compare this to being brought up by a hunter-gatherer. 'He made our country walks so alluring that soon I was enrolled as an apprentice naturalist. Looking back, I see that his simple teaching gave me a feeling for the Earth, for Gaia, which has sustained me.'[2]

Even when they had lived in Brixton, father and son would sometimes go hiking together at the weekends. This became more frequent after they moved to Orpington. Although Tom was now in his mid-fifties, he was short and wiry and could still walk all day, as he had once done for a living. As they trekked across the rolling hills of Kent, Surrey and Sussex, he would tell tales of growing up in the countryside and learning to live off the land. Tom had never gone to school and had learned his alphabet so late in life that he still mouthed words as he read them, but he had picked up a deep, practical knowledge of botany and ethology. He showed Lovelock how to catch rabbits, dig up earthnuts and safely eat the sweet flesh of poisonous yew berries. 'He was that kind of countryman who felt himself to be part of the natural world,' Lovelock recalled. 'For him there were no weeds, pests or vermin; everything alive was, in his view, there for a purpose. He had an immense respect for trees and referred to them as the noblest form of plant life.'[3]

These moments together were precious opportunities to catch up, having been deprived of his parents' company for so long. They left Lovelock with an idyllic impression of the English countryside before it was gored by motorways and stripped by industrial agriculture. This rich landscape formed the soul of his patriotism. He was not alone. The idea of a 'green and pleasant land' evoked by William Blake's 'Jerusalem' had been put to music during the First World War as a morale-boosting call to defend England's nature and culture. Suffragettes like his mother later adopted the song as their anthem. But for Lovelock, the 'green and pleasant land' of rolling meadows and hedgerows would always be associated with Tom. 'I'd walk with Dad whenever I could,' he recalled. 'Up to 1939 you could walk through heathland from Land's End to Brighton.'

Their usual hikes were close to the Pilgrim's Way, which explained his explosive fury when that historic trail was blocked. They would walk the 20-mile stretch from Orpington to Shoreham, passing through the Darent Valley, which was filled with wild orchids. At other times, they would wander further east through Kent, pulling themselves across the River Medway on a rope ferry. Lovelock's most joyful childhood memory was taking a steam train with his father to Holmbury St Mary, ambling through meadows full of buttercups and daisies, and then climbing several hundred feet to the hamlet of Coldharbour. They then enjoyed a long ramble around Leith Hill, stopping for a picnic of sandwiches and ginger beer by the edge of a stream. Liberating him from his mother's worries about 'trade' and 'class', his father made him proud to be descended from country folk. 'He put my hand under the water to tickle trout and we collected bilberries. It was a wonderful day out,' he reminisced.

As would be the case for the rest of his life, the freedom to roam – physically and mentally – was now central to his happiness. This was how he explored the workings of the world, and it was on such walks that he would gather his thoughts. If his father was unable to join him, Lovelock would head out alone, often hiking 40 miles in a day like another of his scientific heroes, Michael Faraday. During his teens, as he grew stronger and more independent, he ventured ever further during the holidays, first a solo bike ride from Cornwall to London, then a week's solitary mountain climbing in Snowdonia. These trips were not escapist. For Lovelock, this was an immersion in reality.

School, on the other hand, was an unwelcome distraction, or what he later described as 'a penalty that children are forced to pay for the crime of being young'. He refused to do all but the most essential homework (arguing it was too exhausting after his long commute) and declined sports entirely ('muddy and a waste of time'). Most of his early Strand-era school reports are filled with complaints about poor attitude and careless disposition. Up until the age of fifteen, there is no hint from his teachers that he might be capable of exceptional thinking. In maths, he continued to struggle. In most other classes, he simply could not be bothered. But there were glimpses of the budding genius.

From his mid teens, Lovelock regularly placed first in chemistry, his favourite subject. He memorised the periodic table and large chunks of the Merck Index of commercial chemicals, and he knew more than many of his teachers about the properties of compounds, especially those that could produce a vile smell or a loud bang. Occasionally, he would even stretch his mind to other fields of study. When he became aware that Tom and Nell were struggling to pay his fees, due to the Depression, Lovelock entered the exam for a scholarship. His form master told him he was wasting his time because his maths was woeful, but he scraped by in arithmetic and scored highly in the essay section by regurgitating extracts from a book on iron and steel that he had recently read and memorised. To his parent's surprise and delight, they never again had to find the money for tuition. For Lovelock, it was a long overdue boost to his self-esteem: 'Just for once, something right had happened.'

In fifth form, he once again astonished his educators when he applied for the school prize – a competition intended for sixth-form students. Once again, he made the rare effort of applying himself and came out top. In announcing the award to his class, the French master spitefully dismissed Lovelock's triumph with a put-down that the teenage boy remembered for the rest of his life: 'This boy isn't intelligent, he is merely a storehouse of knowledge.' Unlike the elite schools of Eton and Harrow, pupils at middle-class Strand were not encouraged to excel or to lead or to challenge, but to know their place. It had been established as a training ground for government functionaries, rather than disruptive thinkers like Lovelock.

'I was an awkward sod. I didn't hesitate to fight,' he recalled. He refused to come into school on Saturdays for extra classes and special events. He would set off stink bombs in the corridor just as the most despised teachers passed by. This gained him the respect of his peers and the nickname 'Mad Scientist'. After years in the social wilderness, he had finally started to make friends. Although Lovelock's physical appearance – small, skinny and bespectacled – might have made him a target for bullies, his sharp wit fended off potential harassers. He knew when to keep his mouth shut, such as during the group masturbation sessions in French class (which, he said, was briefly a vogue for some of the pubescent boys), and when to complain, exemplified by a successful appeal to the headmaster when a teacher threatened canings as group punishment for the misbehaviour of a handful of individuals.

Instead of changing the system, Lovelock's rebellious streak drove him to bypass it. He demonstrated a healthy contempt for the syllabus and focused only on what he was interested in. At the age of twelve, he became fascinated by quantum theory, particle accelerators and the effects of radiation. He read Frederick Soddy's *The Interpretation of Radium* and *Matter and Energy*. His love of fiction extended to the dystopian visions of Aldous Huxley's *Brave New World* and the merciless parody of Stella Gibbons's *Cold Comfort Farm*, in which he no doubt saw echoes of his enforced holidays in the stories of Aunt Doom's family in the village of Howling. But his greatest literary hero of this period was George Orwell. *Down and Out in London and Paris*, *Burmese Days* and *The Road to Wigan Pier* nudged him towards the political left. At school, he would argue the Spanish republican cause against schoolmates who sympathised with Franco's fascists. At home, it made more sense to avoid politics: his mother was a socialist; his father voted Conservative.

These differences never threatened his parents' marriage. The odd couple were, by all accounts, very much in love and would stay that way until Tom's death in 1957. They were close but found space for themselves. Arguments were rare. Nellie would sometimes scold her husband for wasting money. When Tom needed a change of scene, he would wander over to the pub, the bookmakers or take the train into central London for a gathering of the Captain Marryat Fellowship – a literary society based on a series of popular adventure stories about a fictional naval officer – or a catch-up with old colleagues in the Metrogas Slot Collectors' Association. Nobody could understand what he and Nellie saw in one another beyond the canard that opposites attract. As Lovelock remembered them, Nellie was pushy, argumentative and irritated everyone, while Tom was calm, quiet and universally liked. But they found contentment in each other and, now they had retired to the suburbs, they even made a little more time for their son.

Lovelock no longer felt an unwanted outsider in his own home. Even school no longer seemed a prison. His grades started to improve after teachers realised he needed glasses. He was still often chastised in school reports for dodging homework, insufficient effort, failure to take part in outdoor activities and 'shocking' handwriting. But as his teens progressed, his class rankings improved in almost every subject. After the age of fifteen, he was usually first or second not just in chemistry, but physics, history and geography. In 1937, his house master, Medley, expressed hope that Lovelock would pass the higher education examination because his strong science ability would make up for his 'decidedly weak' arithmetic. In his final report in 1938, form master Fowler sends the young man off into the wider world with a prescient observation that innate curiosity would guide him to greater things. 'He seems interested in books and the things around him. This is encouraging.'

During sixth form, when Lovelock felt 'things were finally getting more human', he developed his first close friendship with a boy who was everything he was not. Edward Newton was a brilliant mathematician, talented athlete and head boy. He was also a fascist, who would make up anti-Semitic limericks and eulogise the mobs of Oswald Mosley's blackshirts and their attacks on immigrant communities in the East End. He would tell Lovelock of breaking the windows of Jewish shops and fighting with makeshift weapons, such as a chair leg wrapped in barbed wire. Lovelock said he disapproved of his friend's anti-Semitism, but a tight personal bond developed between them. They would argue for opposite sides in the Spanish Civil War, make up stories together and make fun of their teachers by doodling elaborate caricatures with human faces and worm-like bodies. They called these images 'luvles' (pronounced 'loovals') and Lovelock would continue to draw them for the rest of his life.

There was a homosexual undertone to their intimacy, but Lovelock said it was never openly expressed or made physical. 'He had a pash on me but never tried anything ... I was attracted by his intellectual brilliance. I wasn't good at mathematics, but he could do it in his sleep.' The two friends always sat together in classes but never met outside of school until a year after they had both left when Newton, much to the shock of his family, followed Lovelock into an apprenticeship at a photographic chemical company. During the Second World War, Newton became a decorated RAF pilot and married. His wife refused to let her husband meet Lovelock again.

At school, Newton's friendship had helped to unlock a creative spark in Lovelock. He remained obsessed with science and dreamed only of being a scientist, but he also started to express himself more eloquently in drawing and writing. 'A rough, stony track led me to the wastes of Dartmoor,' started one typical essay on how he had spent the summer holiday. That screed also hinted at Lovelock's future as an ecologist. 'We're proud that England is the most powerful country in the world. I think this is wrong. I think we should be proud to live in the most beautiful country in the world.' The English teacher reportedly told the class the essay showed such promise that Lovelock was the only one among them likely to be published, but he still marked him down because the handwriting 'looked like a drunken spider with its left leg dipped in ink'.

His first published work was 'The Trail of the Lost Asteroid', which appeared in the school magazine in 1937. The villain is a shady professor who is expelled from the Physicist's Club for boring holes into the London Tube and erecting giant mirrors across the track that make drivers go insane thinking they would collide with themselves. All is set to rights when this dastardly character is incinerated in the Deptford Blast Furnace.

Lovelock's most impressive literary work of this period, however, was inspired by the Beckton gasworks. A trip to the giant industrial plant, then one of the biggest in Europe, had been organised by the school's Science Society. Beckton was as close as mid-war twentieth-century Britain came to a high-tech hub. The young chemist was ecstatic to visit the Silicon Valley of his age, not least because his father had worked in the same business. The gasworks covered acres of land, had its own railway and a giant wooden chamber filled with snow-like flakes of narcylene. His favourite part of the trip involved climbing up a ladder and walking across a wooden gangway above a lead-lined chamber filled with sulphuric acid: 'If we had fallen off, it would have been certain death and a very unpleasant one at that. I loved it.' And here too was a reminder of how his father, the former gasworks employee, had re-invented himself after prison. In the impressionable young man's mind, that gave industry a fundamentally benign, transformative power that took on mythic proportions and would never lose its appeal.

The visit inspired a sonnet, a series of sketches and a fictional essay for the school magazine. Written shortly before he left school and possibly in collaboration with Newton, the story, 'Goblin in the Gasworks', conveys the eighteen-year-old's dreams and is remarkably prescient about the direction his life would take.

The tale starts close to home: 'On the damp misty salt marshes of Essex where the dull green sedges merge into the Thames' murky brown, there is spread out in untidy orderliness like the tents and battlemented towers, the turrets and castles of some giant's army camp, the vast bulk of Beckton gasworks.' Lovelock sets this in a time so long past that even 'Old Tom, the road-mender' (an obvious allusion to his father) cannot remember the details, when 'large industrial plants grew wild in riotous profusion and were as yet uncultivated and gardened into the semblance of orderliness that we call present-day civilisation'. He uses himself as the model for the story's teenage protagonist, Nicholas Cobalt, who is a 'kindly, simple soul' of 'exceptional brilliance'. His father, an Essex farmer, works hard to pay for the young lad to be apprenticed into the laboratory of a famous alchemist, who permits him to carry out experiments that were considered by most as 'too difficult and dangerous even to think of' (a prescient description of Jim's later work). With the profits, the young man's first step is to escape into creative isolation (another remarkably accurate glimpse into the writer's future) deep in the countryside. It is near there one night that he rescues an injured 'plant-minding goblin', who is so grateful that the imp shares magical secrets about the deep workings of nature. With this knowledge, Nicholas is able to create 'pipe weeds', 'steam plants' and a natural industrial plant 'capable of supplying all London with gas, and the world with valuable coal tar dyes, medicines and hundreds of other wonderful things'.

Looking back, Lovelock said: 'I revealed myself in that story.' The essay is the culmination of everything he had learned up to that point. It is the product of long hikes through the countryside, solitary scientific enquiry and insights about nature gleaned from his father. It is also the clearest expression of an outlook shaped more by personal exploration and experiment than by the 1930s British curriculum. At this young age, Lovelock's ideas are too immature to be considered a philosophy, but they show a clear intellectual direction, towards science but away from a purely mechanistic explanation of the world. While few other teenagers would consider a lifetime of solitary study to be a fairytale ending, this was Lovelock's vision of contentment. He was engrossed by relationships; they just tended to be between things rather than between people.

There are hints here of the art and holistic vision of his later Gaia theory. Nature and industry are not contrasts but complements, evident in the compound neologisms, such as 'steam plants' and 'pipe weeds,' used to describe the creations that young Nicholas splices together. For a student of his age in a school trained to churn out bureaucrats, this was remarkably original. But, in the wider context of European intellectual traditions, the young essayist was also echoing a bigger and older debate.

Not for the last time, he had put a foot in two opposing philosophical camps. On the one hand, his essay championed the rational Enlightenment laws of physics and chemistry above superstition. On the other, it expressed a Romantic reluctance to accept the dualism between subject and object or between man and nature. It was not necessary for him to have read Kant, Voltaire and Locke, or Goethe, Hugo or Shelley; the two different outlooks were personified in his parents: the Enlightenment reasoning of his upwardly mobile, urban mother versus the occasionally irresponsible Romantic instincts of his laid-back country father. He exhibited elements of both. Much as he preferred his father, he was also indisputably his mother's child.

Of course, the mind of the eighteen-year-old Lovelock was still not fully mature, but his essay reveals several principles that would guide him: A devotion to science and reason wedded to intuition and inspiration. He would put his knowledge at the service of industry, but not at the expense of nature – the two were intermingled. And he would be perfectly happy exploring and experimenting alone, as long as he could share ideas with others and have the freedom to roam.

But perhaps the most prophetic passage in 'Goblin in the Gasworks' was the unusual revelation of how the young man dreamed of ending his days: 'In a little stone house and laboratory deep in the marshes, away from all distracting influences so he could fully concentrate on his work . . . and devote his life to pure scientific research happily ever after.'

Lovelock – the teenage industrialist-naturalist – had set a clear end point for his life. Now he needed the means to get there. This required the acquisition not of money, but knowledge. Peerless for so long, he was beginning to realise he could not take this next step alone.

Helen

'They thought I was killing myself.'

The wartime parties at the National Institute for Medical Research (NIMR) in Hampstead were usually organised on the spur of the moment. The blackout curtains would be drawn across the Gothic window arches, the laboratory benches cleared, and chemical beakers rinsed out and filled with alcohol, which was one of the few items not restricted by rationing. Then, as word spread down the corridor of virologists, pathologists and oncologists, the attendees would start to drift in, removing their white lab coats and loosening their ties and tongues as the men packed their pipes with tobacco and the women sipped on improvised cocktails.

The conversations started on serious topics: work, the Blitz and the likelihood of a German invasion. But the chatter was not all highbrow. As the evening wore on, the discussions would become more ribald and professors would compete in reciting filthy jokes; South African and Canadian doctors would provoke Oxbridge graduates from the home counties about the imminent decline of the British Empire; and all and sundry would share the latest salacious details of institute gossip, much of which centred on the youngest member of this elite gang, James 'Jimmy' Lovelock. How long the party lasted depended on a drinking game. The scientists took it in turns to stand on a table and aim paper clips at a target of concentric circles. Hit the bullseye and you had to down a shot. Miss completely and it was time for everybody to go home.

The institute, a former hospital for consumptives, could easily have been the setting for a horror film. With pepperpot turrets, steeply gabled roofs and stone quoins, it perched on one of the highest northern hills overlooking central London. During wartime, it functioned as a look-out for fire-watchers looking for damage from German bombs and rockets, and, under the aegis of the Medical Research Council (MRC), a workplace for some of the brightest minds in the world of medical science. The director of the institute, Henry Dale, had already won a Nobel Prize for his work on the chemical transmission of nerve impulses. Over the coming two decades, several more members of the MRC would be similarly honoured. These were the pioneers of a golden age for the life sciences in Britain.

Lovelock had finally found a peer group in which he belonged. He was much younger than most of these distinguished scholars and far less qualified, but he had been initiated into an elite, if occasionally sozzled, club: 'They had a nutty society. They made me a member. They all got together at the institute, let their hair down and got pissed. That's how it ended each time.'

At the parties, he invented a cocktail of ether and apple juice, which he proudly – and not entirely convincingly – boasted would make any drinker 'as pissed as a newt in no time, and then twenty minutes later you were stone-cold sober again'. He learned to memorise and recite bawdy limericks, some of which he could still recall eighty years later:

There was a young fellow of Kings
Who grew tired of horses and things.
He said what I desire
Is a boy from the choir
With an arse like a jelly on springs.

Now in his early twenties, Jim was driven as much by hormones as neurons. He was exploring a newfound sexual prowess with a bevy of nurses, which – along with intense work – meant he often did not sleep for days on end. The booze did not help. One time, he arrived at a party with bottles of gin and sherry which he mixed in a half-pint beaker. After imbibing a few glasses, he slipped semi-consciously down the wall and had to be taken home and put to bed. On another occasion, he walked into a lamp-post and had to have stitches in his head, a common mishap in London during blackouts, along with tripping on kerbs and falling down steps. It was funny at first, but then his mentors grew worried. Jim was partying so hard that his work was slipping. These learned friends turned their intellects from the questions of the war to the challenge of fixing their young acolyte up with a steady partner. What those well-intentioned boffins did not know was that Lovelock's heart had recently been broken — by his mother.

Love and sex had been a long time coming in Lovelock's life, but war has a knack of changing everything. When Adolf Hitler's forces invaded Poland in 1939, the twenty-year-old was an aspiring chemist, an avowed pacifist and a reluctant virgin. Up until then, he focused his energy on exploration of the natural world, first at school then as an apprentice for the photographic chemical firm Murray, Bull and Spencer Ltd. Lovelock delighted in this day job, which honed his scientific skills and use of precision instruments, slogging his way through evening classes at Birkbeck College and hiking at the weekends. He was almost always alone, though his desire for a partner was growing.

At a youth hostel in Cumbria in 1939, he developed a romantic attachment to a fellow hiker. Learning that she was studying at Manchester, he enrolled in the university's chemistry department, but the object of his devotion did not reciprocate his feelings, so he was forced to look elsewhere. He learned the best place to meet girls in Manchester was the Ambrose Barlow Catholic Society, which staged regular dance nights in the crypt of the Holy Name Church. Putting aside any religious difference he might have felt as a Quaker, he promptly joined and at the first get-together tentatively approached a young

Irish woman, Mary Delahunty. 'I expected to be rejected. I had very low self-esteem. I was of small stature and wasn't wearing a uniform, but she didn't hold it against me.'

For the first time in his life Lovelock experienced an intimate relationship with a woman. Mary was funny, intelligent, caring and vivacious, fizzing with so much energy and imagination that she could animate any gathering. They shared socialist beliefs, a pacifist horror of the war and an irreverent sense of humour. At the students' union debating society, Lovelock proposed a discussion on the theme of 'God versus Stalin', which was blocked after it came to the ears of an indignant Bishop of Salford. During lectures, which Lovelock found boring, he would pass the time by composing poems, drawing caricatures of his professors and writing missives to his soulmate. Only one remains in the archives, written on students' union paper. It is an invitation for Mary to join a study circle in the cafeteria where 'tea will be drunk and buns eaten to the soothing sound of a hundred scholastic discussions'. Once he gets to the subject of the 'pax' (pacifist) activism they share with Mary's brother Kevin, who was shortly due to face a tribunal over his conscientious objection to the war, the prose crackles with joy, wit and a desire to impress and amuse his love:

Tuesday evening

Dearest Mary Theresa

Gosh, Sunday's meeting was a wonderful show. Kevin was in fighting form, he should be an advocate of a more worthy cause than Ambrose Barlow's little society. But it was his night and he made it. If he can handle his tribunal as well as he handled that meeting, then let us blow all the trumpets of Pax propaganda and declare his trial, when it comes, to be the Saga of Pax and fill the town to overflowing with enthusiastic supporters. Let there be Charabancs driven there from Oswaldtwistle and a procession from Whalley Range, and let us wear rosettes and wave rattles just to show what activism really is.

He adds a piece of doggerel that came to him in that morning's physics lecture with 'Professor Twist':

Of owl I give you now a tale
In starlight dim, in moonlight pale
He sat up high upon his tree
And sang with undiluted glee
The song was old, the words were few
It was of course: To whit! To who!
Through purple night and writhing mist
There passed beneath Professor Twist
He fixed the bird with scornful eye
He paused and spoke, in passing by;
'Desist all night, "to who" to howl;
"To whom" you ungrammatic fowl.'

And then he signs off with: 'Oso many hours of happy thoughtfulness to you. Jim.'

Lovelock was besotted not just by Mary but her entire family, which seemed to provide everything his own lacked. The Delahuntys were Irish migrants whose house in Moss Side was an anarchic hive of activity and affection. As well as Mary's five siblings, there were endless comings and goings of visitors, all of whom brought gossip or songs or scandal to this most open of houses. Mary's widowed mother, who provided for the family with money she earned playing bridge, was famed for her hospitality and cooking, and she happily adopted Jim as one of her own. He would often dine with the family – a welcome break from student canteens – and he and Mary would go on dates to the cinema or music hall with free tickets provided by her uncle, who was a culture critic for the *Manchester Guardian*, or take trips to the seaside at Blackpool, where they caught a first glimpse of the Royal Air Force's new fighter plane, the Supermarine Spitfire.

Other times, they would stay at Mary's home and dance to gramophone records. It was in this house that Lovelock, then twenty years old, and Mary lost their virginity after he was serendipitously obliged to stay overnight due to an air-raid warning. 'We had Hitler to thank for that. I rode my bike to her home. The siren went and there was no all-clear so Mary said: "You had better stay here."' It seemed a natural step. The couple were so obviously, blissfully in love that the assumption of everyone who knew them was that they would marry. Her brother Roger later wrote to Lovelock: 'We all looked at you, somewhat presumptuously, as a future in-law – a prospect wholly acceptable. You fitted into our disorganised-but-very-happy Moss Side household as if "to the manner born".'[1]

But there was one small obstacle to these dreams of marital bliss: the diminutive powerhouse of opprobrium that was Nellie Lovelock. 'It was very clear that Mary was the one I was going to marry,' Lovelock remembered. 'Mum got to hear of it and she exploded. She was very anti-Catholic.' He had grown used to Nellie's relentless disapproval of everything and everyone. The only person who could soothe her was Tom. With everyone else, she was a hackle waiting to be raised. And, of course, no subject made her feel more defensive and indignant than the future of 'her Jim'. Unbeknown to her son, Nellie wrote to Mary to invite her to a London teahouse, where they could talk woman to woman.

The aspiring bride accepted the conditions. She had no wish to upset her prospective mother-in-law and hoped to persuade Nellie to see beyond her anti-Irish and anti-Catholic prejudices. As a gesture of goodwill, she presented her with a bouquet of flowers – no small extravagance on her low income in a wartime era of deprivation. That peace offering was shot to pieces within minutes of the two women sitting down and ordering their beverages. Nellie was contemptuous. 'Why squander money on such a meaningless gesture?' she spat, in what was to prove the opening salvo of a sneering tea-shop tirade. Nellie claimed Mary's Irish Catholicism would jeopardise Jim's career, that she would wreck his concentration with too many children, and warned that she should immediately break off the relationship or all their lives would be made a misery.

Mary was crushed, and she separated from Lovelock without explanation just after Christmas, leaving him heartbroken and confused. Lovelock never forgave Nellie for this sabotage. 'My mother had ambitions for me. She knew I was intelligent and could go far. She didn't want me to marry Mary Delahunty because she thought that she would drag me down. She was completely wrong,' he reflected bitterly. 'I was heartbroken. My stupid, stupid mother.'

Until Britain declared war, Lovelock had made only modest progress towards his ambition to become a scientist. His lab skills had improved at Murray, Bull and Spencer, but his academic development was restrained by a lack of time, low funds and a lack of interest in any type of passive education. He knew he needed qualifications, but he had little patience for lectures. At Manchester, he crashed through the usual three-year undergraduate chemistry BA in less than two years. This haste, compounded by his eternally weak mathematical skills, resulted in a lower-second degree, which would normally have condemned him to a run-of-the-mill laboratory job.

But this was wartime and normal standards no longer applied. Promising talent was now needed more than good grades, and Lovelock clearly possessed an exceptional mind. In an exam on electromagnetic theory at Birkbeck College, he found a novel way to avoid trigonometry by applying the square root of minus one as an operator. For him, it was an act of desperation because he knew his 'arithmetic blindness' would never allow him to find a solution through conventional algebra. When he walked out of the hall halfway through the exam, the other students thought he had given up. But his professor, Patrick Blackett, who would go on to win the Nobel Prize for physics, commended him for the originality of this approach. At Manchester, his laboratory work on sodium bromide ions was so precise that the head of the chemistry department, Alexander Todd, a tall, dark-bearded Scot who would also go on to win a Nobel Prize, initially suspected the student had seen the answers in advance and threatened to discharge him for cheating. Lovelock then replicated his work in front of the professor to demonstrate his accuracy was neither a crib nor a fluke. Todd was so impressed that when Lovelock graduated in 1941, he recommended him to his father-in-law, Sir Henry Dale, the head of the NIMR and yet another Nobel winner.

This was an extraordinary step up for Lovelock. Funded by a penny from every contributor to the national health insurance system, the NIMR was not just Britain's leading medical science organisation, it was arguably also the best in the world. Along with the laboratories in Hampstead, the institute owned a 32-acre farm a few miles away in Mill Hill, which was used to raise laboratory animals. During wartime, the staff had turned from long-term theoretical work to what was known as applied physiology – initially helping the military with deep-sea diving risk assessments, prevention of sea sickness, methods to reduce carbon monoxide poisoning in tanks, and techniques to protect the skin against flames. Later, they would work on vaccines and defence against biological warfare.

The only question about Lovelock's eligibility was whether confidential medical and military information could be trusted in the hands of a Quaker conscientious objector? The interview for the job at the NIMR was conducted by Dale, who asked Lovelock how he would cope with an assignment that was contrary to his conscience. The young applicant fudged the answer, and his ethical qualms, by replying that he would treat each problem on its own merits. This was a turning point. After Lovelock was welcomed into the institute, his first obligation was to sign the Official Secrets Act. From this moment until almost the end of his life, the pacifist Lovelock would put his mind at the service of the country and its military.

The twenty-one-year-old was thrilled. He was now the youngest member of a distinguished faculty and an initiate into a world of confidential, world-class research. He started out as an assistant to Robert Bourdillon, who was pioneering the study of airborne diseases and hygiene with Owen Lidwell. In the early stages of the war, when the German bombing on London was at its height, the primary concern of the institute was to prevent the spread of viruses in crowded, poorly ventilated underground shelters and hospitals.

This was the start of what would become a Lovelock speciality: biochemical tracing. In an early experiment at Guy's Hospital, he and Lidwell would put a chemical marker, such as hydroxyquinoline, on a child's nappy and get a nurse to handle it. They would then put the nurse's uniform under an ultraviolet light to determine which parts were contaminated. Everywhere he went, Lovelock had to measure airborne bacteria using an ingenious instrument developed by Bourdillon called the 'slit sampler'. He would take this cumbersome device, which had to be wheeled around in an infant's pram, to hospitals, airbases and underground shelters, where so many people were crammed together that oxygen levels fell too low to strike a match.

This meant hours every day in the company of medical staff, most of whom were women. Lovelock found himself suddenly in demand because most men his age had been conscripted. 'I picked up a huge group of nurses who were highly sexed and having it off with anyone,' he remembered. There was a string of lovers, including a woman called Cecilia at Guy's and another named Janet at Brompton. Sometimes, they would have sex at the nurses' digs after watching a film or a play in the West End. Other times, they used medical tents inside the Underground station. On one occasion Lovelock enjoyed a quick coupling in a telephone box outside London Bridge station during a blackout. 'Those things happen in wartime. There is a need,' he recalled. In the summer of 1941, Lovelock went without sleep for two or three days a week, and his concentration waned.

Helen

'My friends at the institute were quite worried. They thought I was killing myself.'

When virologist Marinus van den Ende noticed Lovelock was struggling to count samples on a tray, he and another colleague, Frank MacIntosh, hatched a plot to set their protégé back on the straight and narrow. At institute parties and events, they started to nudge Lovelock in the direction of the NIMR receptionist, Helen Hyslop, a 'nice London girl', who was considered smart and steady – the ideal woman, his colleagues thought, to turn his head away from the nurses. They talked up her qualities and encouraged him to take her on dates by providing tickets to the cinema. 'There was continuous pressure on me to be with her . . . They set up things to throw Helen and I together. It was a set-up, an absolute set-up,' Lovelock recalled. 'It was a put-up job by a bunch of top scientists who looked after me like a child. Whenever they wanted something fixed, they would ask me. In return they fixed me up.'

It was hardly love at first sight. He and Helen had been politely greeting each other every morning for more than a year without any particular enthusiasm. Helen, an attractive, slim young blonde of Scottish extraction later told her children she had barely noticed her future husband, who was just another short, bespectacled boffin and a few years her junior. The first time he caught her attention was when he forgot to wear a belt and had to use a tie around his waist to keep his trousers up. The next was when he broke his glasses and she noticed he had nice eyes. Lovelock was similarly lukewarm in the beginning. 'They had to push hard before I would go out with Helen. I wasn't attracted to her at first.'

Their first date was at an Italian restaurant opposite Hampstead Underground station, then Bertorelli's in Soho, followed by Schmidt's in Curzon Street. Helen made it clear that sex should wait until after marriage, so they got by on mutual understanding more than personal chemistry. 'I felt that she felt that if she didn't marry soon she would become an old maid. All the other men were away at the war and she thought I was better than nothing. It was not a very romantic affair,' Lovelock recalled. After a while, though, they developed a rapport and a physical attraction. He may not have felt the spark that had thrilled him with Mary Delahunty, but he was under pressure to do the right thing. He proposed in a pub, the Spaniards Inn near Hampstead Heath: 'I thought, to hell with it. I can't go on chasing nurses all the time. I said: "Helen, let's get married if that's what you want."'

He did not dare to tell his mother until three days before the wedding for fear that she would once again try to sabotage his plans. This proved astute. Nellie was fiercely disapproving, but it was too late to intervene. The ceremony took place just before Christmas at the registry office near Parliament Hill Fields. Jim's best man was David Evans, who developed the whooping cough vaccine. Helen's maid of honour was Marilyn Scott, who was a cook at the institute. Nellie was there too, gritting her teeth and tactlessly complaining to Helen's parents that the marriage would be awful for her son.

The reception was a Bath bun and a pot of tea at Euston station, before the newlyweds headed off to their honeymoon in Keswick on a train full of troops. Lovelock chose the Lake District so he could share his love of the great outdoors with his new wife. Helen struggled gamely with the hiking, but her new husband sometimes had to push her up the fells to ensure they got back to the guest house in time for dinner. Her exhaustion was odd for someone so young. They put it down to a throat infection she had picked up in the cold, damp air.

Decades later, Lovelock looked back at the 'mismatch' with admiration for how the couple managed to make it work: 'I don't think I was in love at any point. I don't think an arranged marriage can be like that. It was totally different from my feelings with Mary Delahunty. In the early twenties one is highly sexed. That made the marriage tolerable. We were both reasonable people. I didn't spend all our money in the pub or gamble.

Helen

She was also sensible, and we both had a sense of humour.' They bought a house in Gayton Crescent, which was within walking distance of the NIMR. Free from landladies, parents and colleagues, they came to enjoy each other's company.

Mutual friends and relatives later expressed surprise at claims that this was a loveless match, as Jim stated in his memoirs and late-life interviews. The couple shared jokes and ideas. Helen was a sounding board, a font of common sense and protector of her husband, who was not always sure whom he could trust. They were openly affectionate, by the restrained standards of the time. A letter Lovelock wrote to Helen from Grafton Underwood airbase sings with tenderness, even though it is written in comic imitation of the US pilots he was working with at the time.

15 October 1943

Helen darling
Here is your loving husband again, pushing the piddling pen just for to charm his honey. I can tell you before writing more that this letter will be lousy. I'm so full of grub and so worn out that thinking is just impossible. Still, the only thing I can think of is coming home to you. You know in a joint like this, where all the fellas are thousands of miles from home, writing home seems a mighty important thing, and going home like the ending of a prison sentence. It makes me realise one hell of a lot darling what a lucky bugger am I. I often worry when I am away that you and Betty might become girlfriends; that's a terrible thought but there's no limit to the lengths that lonely women go!!! But always I take your picture out of my wallet and gaze into those dear strong eyes and know that you will be faithful always.

Well my sweet I guess I must can the crap and say cheerio for now

All my love and much more. Yours forever. Jim

Their first daughter, Christine, was born in New End Hospital on Highgate Road on 16 September 1944, a week after the first V-2 rocket hit London. Her father later joked that this made her one of the first children of the space age. She was a petite baby, just 4 pounds (1.8 kilograms) – the result perhaps of wartime rationing for her mother. Lovelock was worried. He had not had a parental model in his infancy, and he was not sure how to cope: 'I didn't have strong fatherhood feelings. I was just concerned about her small size. But she put on weight very soon with extra rations of milk and vitamins.' A second daughter, Jane, followed on 26 February 1946.

Helen proved a calming presence and, as would be the case for most of the next four decades, provided secretarial support and a solid home base from which Lovelock could embark on his scientific adventures. She did not want her patriotic family to know her husband was a conscientious objector. 'Poor Helen. She was so afraid of what her working-class neighbours would say,' Jim recalled. 'I had to leave the Quaker church because she was anti-religion.'[2]

There were other reasons Lovelock may not have wanted to remain a Quaker at this time. His colleagues had convinced him of the righteousness of the fight against Nazi Germany, leading him to become more involved with research that was part of the war effort. He persuaded himself this did not contradict his pacifist principles: 'Our prime objective was what medical services do generally, they're not out to kill people, they're out to care . . . With a Quaker background it would have been very difficult for me to have been involved on the other side of the research. I don't think I would have wanted to do it.' This was somewhat disingenuous. He would have been well aware that much of the science he was involved in was dual use and had a military application.

When Britain was on the defensive in the early stage of the war, Lovelock was part of a team that assessed the Petroleum Warfare Department's proposal to discharge and then ignite vast quantities of oil in the Channel to deter a German invasion.

Winston Churchill was enthusiastic about the plan and had approved a trial at Studland Bay in Dorset. 'It was a ridiculous idea,' Lovelock recalled, laughing. 'We showed that even a wooden boat could pass through the flames unscathed.' Later, he joined troops testing flamethrowers in bombed-out areas of Canning Town. His job was to measure the heat and how best to protect the human body. He discovered the simplest form of resistance against fire was a woollen blanket. Years later, when he started to think about Gaian interactions between life and the environment, he realised mammals evolve defence mechanisms against forest fires. Fur and wool do not burn; they turn into a layer of carbonised matter which is non-flammable and has almost zero heat conduction, so the skin underneath is insulated.

Lovelock was asked to come up with a method to determine whether a burn victim's injuries were first, second or third degree. He had been allocated several shaved rabbits for this experiment, but he had other ideas. He guessed mustard gas paper, which had been developed in the First World War, could be adapted for this task, because it turned three different colours depending on the chemical environment. Rather than bother the poor rabbits, he tested his theory on himself by holding a Bunsen burner to his flesh long enough to destroy the epidermis and dermis and to leave his skin blistered and slightly charred. The test paper worked, but his third-degree injuries had to be treated. The doctor who tended them was Frank Hawking, who later invited him home for a meal where his wife, Isobel, asked Jim to hold her baby while she prepared the food. Lovelock and that infant, the future cosmologist Stephen Hawking, would not meet again until 1974 when they were both made fellows of the Royal Society.

The end of the war was now in sight and the clear winner was science. Radar technology, jet engines, electro-mechanical computers, blood transfusions, superglue, synthetic rubber, penicillin and duct tape were all invented or rushed to industrial scale by the exigencies of the war. It had also ushered in an era of terrifying new superweapons. Wernher von Braun's V-2 rockets, which would later propel a race for space, were hitting London, though it was obvious to everyone at the institute that these powerful but imprecise weapons were a sign of Germany's desperation rather than its strength. The talk in the canteen was already turning to the battle against Japan in the Pacific and a powerful new weapon that the Allies were developing. The A-bomb was supposed to be top secret, but Lovelock said the atomic research establishment leaked like a sieve because there was an overlap of personnel: 'I knew about the possibility of an atomic bomb by about 1941. There were people interfacing the atomic side and the medical side. So we knew what was going on. It was quite clear that somebody was making a bomb.'

When the nuclear weapon was first used against Hiroshima and Nagasaki, the Quaker in him was horrified, but the bomb-maker was fascinated. 'I could not help but wonder how it worked. I didn't expect it to be used against civilians. That was unforgivable. It could have been dropped on an island near Japan and had a similar effect with fewer casualties.' Nuclear war was now a reality, and Lovelock would frequently be on the fringes of the technology that made it a possibility. His friend Frank MacIntosh later served as an official observer at the atomic bomb trials at Bikini Atoll, where he applied the three-colour strips Lovelock had developed to estimate burn intensities. Lovelock claimed he was astonished to learn of this: 'I had no idea. I was unbelievably innocent.' Over the following decades, however, he too often worked with radioactive materials and, in the 1980s, he conducted a large-scale test of the likely distribution of fallout from a nuclear explosion.

War had indeed changed everything. By the end of the global conflict in 1945, the pacifist virgin of six years earlier had become a husband and father who willingly participated in what he saw as a morally justifiable conflict. He was still determined not to hurt anyone directly, but, knowingly or unknowingly, he had, like the vast majority of senior scientists of his generation, started to work on dual-use technologies, often shrouded in secrecy. This was what war demanded. It was what the British state needed. It was the price he was more than willing to pay to be part of the elite scientific club that had welcomed him as one of its own. Lovelock had learned from some of the world's leading biochemists how to read the air, to detect and follow bacteria and viruses. On the cusp of the Cold War era, this was an invaluable skill, though no longer one that could be honed in a population centre like London. He, Helen and their young family would need to move somewhere altogether quieter and more remote.

Christine, Jane, Andrew, John

'The most lovely place to grow up.'

For the infant Christine, it felt like a holiday camp. Each morning, she woke up in a cosy, centrally heated chalet on a hillside overlooking the Wiltshire Downs. Mummy and Daddy would already be awake, drinking coffee and reading the morning newspapers. They got on well and often held hands and kissed one another. Mummy never needed to worry about cooking because meals were delivered four times a day. Christine loved watching the food – kept warm in plates covered by metal thermos lids – being pushed in trolleys along the covered boardwalks between the kitchen and the huts. In an era when other adults grumbled about rationing, her family seemed fortunate.

And what a marvellous job her daddy had, just a short walk away in the laboratory building where he put on a white coat and did awfully clever things to understand why people caught colds. Sometimes, if she or her sister, Jane, had a runny nose or a sore throat, they would be asked to help. Daddy would organise infection parties where they played 'pass the parcel' and 'pass the balloon under the neck' with volunteers who strangely wanted to catch their germs. It was fun. But Christine was shy, which was a handicap when it came to passing on bugs. Jane, on the other hand, was very sociable. Daddy joked that Jane was a super-spreader. Most of the time, they would just run wild with the other children who lived in the camp, wandering the nearby fields and playing games in the woods. Mummy and Daddy never seemed to mind what they got up to, as long as they were back in time for the next meal. Daddy said that was how he had grown up and it didn't do him any harm. What dangers could there be, anyway? There was nobody else for miles around. It really was the most lovely place to grow up.

This 'lovely place' was the Common Cold Research Unit (CCRU), a former military hospital perched on a hilltop surrounded by open fields a few miles outside of Salisbury. The US military had donated this site, known as Harvard Hospital, to the British government in the immediate aftermath of the war on condition that it be used for the study of respiratory diseases. The cluster of Nissen huts was put under the jurisdiction of the MRC and staffed with Britain's leading experts on virology and air hygiene. Lovelock and his family moved in almost as soon as the facility opened in 1946. On the face of it, this was a bemusing career step for the aspiring scientist. From the intellectual hothouse of Hampstead and a victorious war effort, he had chosen the fringes of Wiltshire and a multi-year examination of how people develop snottiness, sneezes and sniffles. Nobody doubted the common cold was important, but humanity had lived with this mostly non-lethal ailment from time immemorial. There were surely more urgent matters to look into.

Lovelock said he was motivated by a mix of pragmatism and romance. The new job came with a 50 per cent pay rise, accommodation and free food. It gave him the chance to work under one of the brightest stars in medical science: Christopher Andrewes, the Nobel-winning discoverer of the influenza virus. Best of all was the chance to live in an area of the countryside close to Bowerchalke, a Wiltshire village that Lovelock had fallen in love with ten years earlier on a bike ride across southern England. This was the landscape his father had encouraged him to revere. It was also the first of many retreats into isolation and nature that enabled him to work on sensitive projects far from prying eyes.

The timing and the location of the move were certainly curious. Harvard Hospital was less than 10 miles from Porton Down, the UK's top-secret military research establishment and the centre of the country's chemical and biological weapons programme, which was being ramped up to develop a deterrent against a possible Soviet germ attack. The Ministry of Defence (MoD) has always insisted the CCRU and Porton Down were separate. However, towards the end of his life, Lovelock believed there was an institutional connection between the two establishments. The staff would mix together. Lovelock visited the laboratories at Porton Down several times and became close friends with Thomas Nash, a chemist who would later work there. Technology was also shared by the two facilities. Both, for example, used a Porton Impinger, which was a device for sampling airborne viruses or bacteria. This does not necessarily mean the CCRU conducted dangerous-weapons experiments, but it could easily have had a symbiotic dual-use relationship with Porton Down, offering insights into the common cold while simultaneously providing a stream of human volunteers on which to conduct non-lethal experiments on contagion pathways and defences.

'It was probably the only place in the world where human beings were deliberately infected with respiratory viruses,' a former head of the CCRU wrote later.[1] Many of the human guinea pigs who participated were from the armed forces or other forms of government employment. To attract volunteers, the CCRU offered free food and accommodation, recreation facilities and paid time off work. Compared to the deprivation in post-war Britain, the lifestyle was comfortable. Bottles of milk and a newspaper were dropped outside the huts each morning and the typical daily menu was porridge, boiled eggs, bread, butter and marmalade for breakfast; soup, corned beef, cabbage and potatoes for lunch; roast chicken, peas, potatoes, gravy, pears, peaches and cream for dinner; and bread, butter, jam and tarts for a pre-bedtime supper. Volunteers wrote to thank the organisers for the relaxing holiday in the country and compared the experience to a Butlin's holiday camp.

These were happy days for the Lovelocks. They were close, lived healthily and could afford a first car, though Lovelock failed his driving test three times before passing. 'When we were growing up, we thought Dad was the best driver in the world,' Christine recalled. 'So we were surprised when friends told us they were afraid to get in the car if Dad was driving. It wasn't just that he drove fast, it was that he didn't know left from right.' Later in life, she had the same problem. Much as he enjoyed hurtling his Morris along winding country roads, Lovelock's favourite pastime remained wandering on foot through the Downs, often with his family.

He would enthral his kids with seemingly magical inventions, including the then great novelty of television. TV broadcasts had been halted for the duration of the war due to fears that German bombers would use the powerful transmissions from Alexandra Palace as a navigation aid. After services resumed in 1946, Lovelock built his own set from army surplus valves and radar equipment. All that was missing was a tall antenna. Too impatient to dig the foundations for such a tower, he used nitroglycerin to do the work for him, much to the delight of his daughters and the dismay of his wife when the explosion left cracks on the walls of the family hut.

Later, Lovelock's idea of diverting his baby son, Andrew, who had been born on 2 November 1951, was to fill empty milk tins with gas and then ignite them with a boom. 'You never knew what was going to happen with Dad. One time, he tied a hydrogen balloon to Jane's toy pram and it floated off,' Christine said. 'Dad was always coming up with big ideas and funny stories.' Their father inculcated the same sense of mischief in his daughters. Whenever they posed for a picture outside the Sunday school building, he encouraged them to lift up their skirts and pull a funny face.

Jane was closer to her mother and less enthralled by her father's unorthodox parenting, but she appreciated his sense of mischief: 'He wasn't a boring old fart. His boundaries of what is permissible were different. My sister and I got used to it. He taught us how to make volcanoes. We'd dig a hole in the garden, then fill it with sugar and weed killer, and shape it into a cone. Then, when you lit it up, it erupted. You've got to get dirty and dangerous to have fun.' She relished the moments when he shared his knowledge of the countryside, passing on what he had learned from his father. 'It was wonderful to go on walks with him as he'd talk about botany, show us bees and orchards.'

The idyll was broken by visits from Tom and Nellie, who managed to infuriate everyone. When they went on a stroll together, Nellie would insist on holding her son's hand while her daughter-in-law walked behind with Tom. When they went for a drive, she would expect to sit in the front. 'She always put her foot in it,' Christine remembered. 'One time, Grandma had an argument with Christopher Andrewes during a meal at Harvard. After that, Dad was told never to bring his mother to the mess again.'

Under Andrewes's tutelage at Harvard Hospital, Lovelock was further honing his tracing techniques and widening his understanding of human physiology, which would later underpin the view posited by the Gaia hypothesis that the world is a living organism. In 1947, it yielded the most impressive inventions of Lovelock's life up to that point, including an ionising anemometer capable of measuring slow-moving air currents, more commonly known as draughts.

The device used radium and mesothorium, from dial paint that Lovelock had scraped off war-damaged aircraft gauges and then electroplated with polonium to 'label' ions with a positive charge so they could be easily tracked. This use of a radioactive source was a major evolution of his tracing skills and was a precursor of more sensitive devices that enabled him to grasp the biochemical dynamics of the planet's atmosphere. He was enthusiastic about the potential uses of radiation, which was then still a new field: 'We were pioneers. Radioactivity was nowhere near as bad as people make out.' Later, the MRC asked him to find a radioactive source that could sterilise rats. He wrote to the government's radiopharmaceutical suppliers at Amersham for a bottle of gold chloride, which gave off X-rays. He would then stand behind a thick pane of glass, open the bottle with long tongs, expose the rat, and then put the closed bottle back in a lead container. 'We were not in the least afraid,' he remembered.

In retrospect, Lovelock acknowledged the work was sometimes dangerous. At one point, he used thorium isotope and a solution of uranium-238, until he was warned these materials could cause health problems. When he requested a safer element for his tracers, government suppliers sent him a box of polonium-plated material in the next day's mail. This isotope can be so deadly if inhaled or swallowed that it was later used as an assassination tool by the KGB. Lovelock knew radiation could get out of hand. On a later occasion, when he was working on iodine tracing in Mill Hill, all the radioactive readings suddenly went off the charts. It turned out there had been a fire at the Windscale nuclear plant that had created a plume of uranium-235, which drifted across the south of England.

He felt these risks were tolerable, when balanced against the benefits of radiopharmaceuticals and nuclear power. He was proud of his reputation as someone who could handle radioactive materials and he was later offered the chance to head the Radiation Biology Unit for the Atomic Energy Research Establishment. This helps to explain his passionate advocacy of nuclear power, which decades later caused his deep rupture with the green movement. He was a lifelong critic of health and safety rules and would make a profitable sideline out of taking on dangerous, off-the-books research, and not just in handling explosives and radioactive materials.

While working at the CCRU, Lovelock published two papers in the *British Medical Journal* on the transfer of infections through handkerchiefs and how to disinfect handkerchiefs. These apparently banal topics belied the specialism of his co-author, Keith Dumbell, a virologist who would later work at Porton Down and be revealed as the indirect source of a smallpox outbreak that killed one woman and caused a medical health emergency in Birmingham in 1978.[2] Lovelock said Dumbell illegally kept samples of smallpox and anthrax at the CCRU in the late 1940s and early 1950s so that he could do off-the-book tests.

Lovelock almost fell victim. He cut his knuckle on a sliver of laboratory glass that he claimed had accidentally become embedded in a bench at Harvard Hospital. He cleaned the wound with antiseptic, but an itchy, milky white blister developed with a black blob in the centre surrounded by a transparent ring of pus. Soon after, he felt feverish. At first, he thought he had the flu, then red lines stretched up the lymphatic channels in his arms and he drifted in and out of consciousness for almost a week, while Helen cared for him. A tropical medicine specialist at the facility jabbed his backside with a huge injection of penicillin and asked Lovelock whether he had been working on 'something nasty'. Many days later, after he had recovered, Jim heard from his colleague Dr J. Porterfield and from a technician that Dumbell had been trying to grow anthrax spores on an agar plate that had splintered. When he took this allegation to the head of the hospital, he was told his claims were 'impossible'.

Decades later, Lovelock developed a clearer theory of what had happened. By then, he too had worked at Porton Down at the 'interesting science' section of the military research facility and spent many years as a consultant to the MoD, carrying out experiments on radioactive and explosive materials that the government could never admit to doing in-house. Based on this experience, he said he could understand how Dumbell, who was initially an expert on tuberculosis, came to work in a freelance capacity with illegal pathogens. 'I think Dumbell's relationship with the security services was the same as mine. We were outlying scientists who could be given dangerous jobs.'

Lovelock also admitted that he got a thrill from doing risky experiments 'like getting near nasty viruses. I'd grown up thinking you've got to be brave or you are not a worthy person. It was all a bit childish. Really it is. But that's how it was. A really good psychologist would tell you what is behind it all.'

Was Lovelock knowingly involved in biological weapons development? His studies on air movement could have been used for both the containment or dispersal of pathogens. In 1949, Lovelock volunteered to join the light aircraft carrier HMS *Vengeance* for a six-week experimental voyage to the Arctic Circle, where his role was to examine air ventilation in the mess and sleeping quarters using his anemometer. Coincidentally or not, the Royal Navy had the previous year been given responsibility for field-testing biological weapons developed at Porton Down in remote areas, such as the waters around the Outer Hebrides and the Caribbean. Regardless of whether Lovelock was aware of this, his research on air movements within ships could have been useful for those trials.

Science and the military had become intertwined during the two world wars, and they were growing even closer in the Cold War. There was no clear boundary, certainly not as far as Lovelock was concerned. His ethical position was informed on the one hand by his Quaker pacifism, but also by a fascination with explosives and a patriotic desire to support his country: 'I never stopped being a pacifist. I felt World War Two was unavoidable. The pacifism was something that I thought my mum had put me into. My position was that in no circumstance would I kill or knowingly harm anyone, but I wasn't afraid of danger. It's the same now. I can't say for sure, though, because you don't know the consequences of what you do in war.' This gave him a remit to work on dual-use technologies and to offer indirect support for arms development, the atomic and hydrogen bomb programmes and target tracking for the intelligence services. By his own account, he would continue to be involved with the MoD on and off until he was 100 years old.

It did not seem to matter that his research on the common cold petered out without making any discernible progress. 'All we found out after five years was that the common cold had nothing to do with being cold,' Lovelock said. But the lack of published results was no impediment to his career: when he returned to London in 1951, he was welcomed into the senior echelons of the MRC.

Much to the children's irritation, the family spent most of the 1950s in the London commuter belt suburb of Finchley. After the central heating at Harvard Hospital, their semi-detached home in Westbury Road was bitterly cold in the winter. Instead of running free across the countryside, they had to go to school. The transition was tough. The sisters showed signs of inheriting their father's dyscalculia. 'I was useless at school and failed everything. I couldn't tell the difference between eighteen and eighty-one. It wasn't until my fifties that I could tell left from right,' Jane recalled. Christine said she was a year ahead in almost every subject, but she had difficulty with maths: 'It was a kind of incomprehension, as if there was some key I was missing. I am still hopeless at giving change, especially if someone talks to me at the same time, and I can't tell left or right.' Andrew, on the other hand, was exceptional with numbers.

Jane was sociable and enjoyed school, but she struggled to adjust. Born with a divergent squint that her parents ignored for years, she would often knock things over, irritating her father with her apparent clumsiness. She too showed dyslexic traits and would write her name backwards. Her father and mother were hardly the types to push her to get better grades. 'My schooling was weak. I was virtually uneducated,' remembered Jane. Compared to the staid parents of her friends, Jane started to comprehend that her father and mother were unusually colourful characters. Helen painted the house bright green, the woodwork white and the front door pillarbox red. For Jane's bedroom she went a step further, daubing the walls baby blue and the skirting boards in bright pink. She would sing, not well but loudly. 'Mum was as eccentric as Dad, in her own way,' she recalled. The couple were, in Christine's words, 'all over each other'.

The children were later shocked by Lovelock's claim in his memoirs that his marriage was a disaster from the beginning. 'They were very loving to each other when I was growing up,' Jane recalled. 'As a child, every night he'd come home and they'd hold hands and have tea in the sitting room.' Andrew echoed this sentiment: 'I always thought they were the ideal couple. They really had fun together. Mum was funny and kind. She kept him grounded.' Their mother enjoyed an evening tipple of Tio Pepe Sherry, after which she and Lovelock would watch TV shows together. She did a lot of knitting and sewing and would make frocks for her daughters. Her real passion was gardening, and the children recall her looking happiest with a trowel in her hands.

Health was a concern. Helen was laid up from time to time by a mystery ailment that her male doctors misdiagnosed as a womb infection. Lovelock did not know how to cope with her sickness, and this would make him irritated, which upset the children. To soothe them, Helen would say their father didn't mean it; he was frightened because he depended on her so much, emotionally, and couldn't help himself. Another way to look at this behaviour, which was to recur throughout Lovelock's life, is that he struggled to address problems that could not be solved. Vulnerability made him uncomfortable.

On 20 February 1956, Helen gave birth to her fourth child at Great Ormond Street children's hospital. John was born with an undeveloped oesophagus. Half-strangled by the umbilical cord, he needed emergency surgery. He survived but suffered brain damage and was later classified, in the language of the time, as a 'spastic'. He experienced learning disabilities for the rest of his life, though he also lived longer and achieved far more than anyone believed possible in his early days. Growing up, his hyperactivity, autism and wandering spirit made him a 'handful', as his parents would euphemistically say. He was fascinated by extreme weather and loved to go out in fierce gales and biting rain. 'He was a boy who wanted to catch the wind,' recalled Christine. Like his father, he had an extraordinary memory and could recall the precise dates of every storm for decades afterwards. 'I think he was more like Dad than any of us,' she said.

In Finchley, family life revolved around Lovelock, whose word was law. Everyone recognised he possessed special gifts. 'We were like disciples. Mum thought he was amazing. We all thought he was clever,' Christine recalled. 'Talking with Dad was like having Wikipedia or Google long before they existed. Wherever we went, he knew everything. He was like an encyclopedia. We knew he was a genius. Everyone knew. Dad was not like a normal father, he was a genius father.' Jane was less impressed: 'I always wanted a father more than a scientist. I was never a god worshipper. I wanted a dad.'

Without a parenting model of his own, Lovelock was never entirely comfortable in this role. He was far surer of himself as an inventor. In the 1950s, Jim was starting to make a name for himself among some of the greatest life scientists of the era. Under the directorship of Charles Harington, the MRC had moved to Mill Hill and established itself as a hive of world-class innovation. Every floor had a laboratory led by a Nobel Prize winner: Rodney Porter, who became a laureate for identifying proteins; John Cornforth, who was recognised for his work on steroid molecules; and Archer Martin, who won the prize for developing gas partition chromatography, a technique to separate and measure different compounds, which paved the way for research into fatty acids, the lipid metabolism of the blood and the nutritional value of food. Others forged new techniques on the control of influenza, the low temperature preservation of cells, interferon control of hypertension, the isolation of calciferol and the synthesis of trypanocidal drugs. Mill Hill scientists were encouraged to be creative and to collaborate outside of their fields, creating the perfect environment for Lovelock, who was frequently called upon to build instruments or find novel approaches to lab challenges. 'I had a reputation: "If anything goes wrong, go and see Jim."'

His laboratory assistant at the time, Peter Simmonds, recalled how colleagues who found themselves stuck with seemingly intractable problems would have a cup of tea with Lovelock and chat their way to a solution. 'With Jim there were always ten ideas; nine of them would be outlandish and impossible but there was always one gem.' Simmonds, who later followed his boss to the United States, said Lovelock's creativity and humour stood out. Lovelock would also amuse himself by sucking helium and imitating the squeaky voice of Tweetie Pie. By the mid 1950s, he was a world-class chemist, a biologist, a virologist and a cryobiologist. 'I've never met a scientist who covered so much ground,' Simmons said. 'Mill Hill produced more Nobel laureates than anywhere else at the time, but Jim was an absolute maverick. The other scientists didn't have the spark that he had.'

It was an era of pipes and cigarettes, crossword puzzles and traipsing daily from one part of the North London commuter-belt to another. Lovelock would sometimes feel bored by this dull routine and the often trivial tasks he would be set, but it coincided with a prolific period of invention that opened doors and provided insights for decades to come. He was proving himself a true polymath, a scientist who could turn his hand – and his mind – to just about anything.

In the field of livestock breeding, he and his colleagues made key breakthroughs on sperm[3] and ovary freezing, which meant it was no longer necessary to send prize bulls across the world. He examined the links between coronary disease and fatty acids, and he was funded by Kraft, the food producer, to investigate whether sugar makes people fat,[4] which contributed to the healthy-eating revolution of subsequent decades. The Ministry of Agriculture, Fisheries and Food called on him to fit a radio transmitter onto a bull in one of the first operations to remotely monitor cattle movements. With the great cryobiologist Audrey Smith, he developed a technique to freeze live hamsters so thoroughly they became lumps of ice that could be banged on the table, only to re-animate them later in a hand-crafted microwave heater.[5] This was invented by Jim from parts taken from surplus army radar equipment because he wanted to reduce the suffering of the hamsters, who were previously defrosted with hot metal, which left burns. He later used the device to cook a potato, prompting him to speculate he may have developed one of the first small-scale microwave ovens,[6] though the technology had been around since the end of the war.

Lovelock's reputation was growing. For six weeks, he was scientific adviser on a BBC drama, *The Critical Point*, about freezing people, which was based on the work done at Mill Hill. He also made a sound-effect tape and gave tips to the BBC that, he claimed, inspired the corporation to set up their own radiophonic workshop.

This was also the period when Lovelock made the most consequential invention of his life – an instrument that allowed him to measure the composition of gases and liquids with a degree of sensitivity that no other scientist on Earth possessed at the time. He did not fully understand how it worked until decades later, when he would talk in hallowed terms of his device's 'quantum' properties and 'magical' effectiveness. Without it, his career would have taken a very different path and he would never have learned to scrutinise the sky in such exquisite detail. That instrument was the electron capture detector (ECD).

It started in 1956 with the usual morning coffee ritual at Mill Hill. Every day, a group of elite scientists, including Tony James, Rosalind Pitt-Rivers, George Popjak, Joan Webb and Lovelock would meet in the lab of Archer Martin, the world's leading gas chromatography expert. They would drink coffee and smoke copiously while they solved the *Times* crossword. Nobody was allowed to leave and start work until the puzzle was finished. It usually took ten minutes. When that was done and the others had left, Lovelock asked Martin whether he could analyse the blood of hamsters to understand why they alone – of all laboratory mammals – could successfully be reanimated after freezing. The answer came in the form of a challenge. Martin said the sample was too small for his instruments, so he could only examine the rodent if Lovelock invented a more sensitive detector.

The gauntlet thrown down, Lovelock retreated to his lab and remembered the strange behaviour of his ionising anemometer on the HMS *Vengeance*, when cigarette smoke and other chemical compounds had produced a powerful response. Did this, he wondered, suggest the ionising technique could be used to distinguish not just the speed of the air, but also its composition? He also knew of an earlier ionising detector developed at Shell Development in the Netherlands by Hendrik Boer, who had visited Mill Hill the previous year. Jim ordered a range of radioactive materials from the Atomic Energy Research Establishment in Harwell with which to test his theories and set about combining them with different containment vessels and carrier gases.

Over the following few years, he returned to Martin with not one, but two ionising detectors that were smaller, cheaper and hundreds of times more sensitive than anything that had existed before. The first, which used argon (an accidental discovery because the institute store had run out of nitrogen), proved an immediate commercial success and was later snapped up by the scientific instruments manufacturer W.G. Pye and used by the petroleum industry to produce high-octane gasoline. The second, which used nitrogen as a carrier gas passed through a detector plated with strontium-90 (later replaced by nickel-63), was the ECD. This was a historic breakthrough, though the initial results were so dramatic that the instrument was unable to function for a week and almost written off as too unstable.

Lovelock felt there was something mystical about the instrument because it had an uncanny knack of responding dramatically to carcinogenics and other toxic substances. With refinement over the years, the device became increasingly sensitive. The first working model in 1962 could measure parts per billion, a 1,000-fold improvement over anything that had come before. This was described by one chemist as the equivalent of 'establishing that the extreme dryness of a particularly good Martini is due to the fact that only a single drop of vermouth was added to 125,000 gallons of quality gin'.[7] Later, this was further enhanced to the point where the ECD could detect parts per trillion — the equivalent of a single drop of water in a dozen Olympic swimming pools or a needle in a hillside full of haystacks.

This degree of sensitivity would transform the way humanity sees the world because, for the first time, it was possible to detect the build-up of tiny man-made substances in the vastest and most remote locations. The ECD later revealed the presence of ozone-depleting gases in the stratosphere and carcinogenic compounds in the air, soil and water. It identified the source of summer haze as man-made pollution rather than a natural phenomenon. It was used to detect hidden explosives such as nitroglycerin, TNT and Semtex. And it was tested as a form of intercontinental code-signalling whereby a secret agent could tip a bottle of uniquely labelled perfluorocarbon on the floor in China that could then be picked up one or two weeks later in the UK. Most importantly for Lovelock, the ECD was the key that unlocked Gaia's secrets. Without it he would never have discovered the iodide and sulphur cycles, which are an essential part of the planet's metabolism.

These inventions changed Lovelock's life. As word of his devices spread, he suddenly found himself in huge demand. Shell and other British corporations asked for demonstrations, and US academic institutions, including the Environmental Protection Agency, followed soon after. At the turn of the decade, Lovelock was flitting between the MRC in London and fellowships in the United States, first at Boston, then the University of Houston. On three occasions, he took his family to the US with him for more than a year. They would cross the Atlantic on ocean liners, find a home and then Helen and the children would try to settle while Lovelock jetted around the country. In the space of two years, he wrote home from Tennessee, Pittsburgh, Boston, Los Angeles, New York, San Francisco, Yale, Washington, Buffalo, Chicago, Cincinnati, Michigan, New Haven and countless smaller towns. He travelled light, with a suitcase containing his toothbrush and a change of clothes and a small case with a demonstration model of his chromatograph. 'My arrival with a briefcase became a legend,' he recalled. 'Hordes of chemists came flooding in. I went to every small town to give lectures on detectors of gas chromatography because they were all starting to use them for everything.'

Martin had innovated the technique and Lovelock had refined it. Now, dieticians could distinguish between saturated and unsaturated fats. Environmental agencies could more easily detect the build-up of tiny man-made chemicals in the rivers and earth. But the big-money use was the petrochemical industry. For the first time, oil firms could identify the different compounds in petrol and diesel, and modify the octane number of fuel rather than sell it all as regular. 'They made billions and billions from this. They didn't know how to do it before my device,' Jim boasted.

The Lovelocks were starting to enjoy a more comfortable lifestyle, thanks to increasingly generous stipends from US research trips. On their first visit to the US in 1954 — before the gas chromatography revolution — they had relied on an inadequate Rockefeller fellowship to Harvard Medical School that barely covered the rent, so the family had to burn items of furniture to heat the house and Lovelock sold his blood for $50 a pint to make ends meet. When they moved back to the UK, Lovelock had to enter and win a science-writing competition so he could pay for the passage home with the prize money.

By the time of their return to the United States in 1958, he was a far more valuable asset because of his expertise in chromatography. Yale University funded him to conduct development work on the argon detector and the ECD. The patents should have been worth a fortune, but, not for the first or last time, Lovelock was more interested in pursuing science than cashing in. His host institution claimed rights on the technology. The MRC ought to have had a bigger claim, but British science was still something of a gentlemen's club that looked at commercialisation as grubby. Lovelock said he did not make a penny from this initial patent: 'Many scientists and especially those at Mill Hill were motivated by a simple idealism. Romantically, we imagined that our sole duty was the good of the public and that it would be morally wrong to profit from our research.'

On the Lovelocks' return to the UK, Jane said her father felt jaded by his tenured civil service job: 'Dad was fed up at Mill Hill and wanted to do something different. He felt he could go in, do the crossword and nobody would notice if he did that until retirement. It was very stultifying.'

He also felt hemmed in by his mother. Their relationship had turned on its head in 1957 with the death of Tom. Until that moment, Nellie rarely visited her son's family, and when she did, mild-mannered Tom had been there to smooth any ruffled feathers. When he passed, the floor of her life collapsed. She was suddenly alone. From that moment on, Nellie began to crave her son's attention. At the funeral, her sisters told Jim that he must now step up and be the man in his mother's life. It was an enormous obligation.

Nellie rented a flat close to Jim's home in Finchley and used to drop in at least once a week. The family came to dread her visits. 'It was a trauma whenever she turned up. We had to bite our tongues until they bled,' Jane remembered. 'After she came to visit, we were wrecked. Dad didn't know how to handle her. It was a tortuous relationship with guilt on both sides. I felt sorry for her. She was a tortured soul in many ways. She wasn't maternal at all. Dad didn't have good parenting. He was crippled.'

The arguments, which are grimly documented in Nellie's diaries, grew so traumatic that the family plotted an escape. On 11 May 1960, they moved to the remote village of Bowerchalke in Wiltshire, where the family had previously rented a holiday cottage. It was a huge gamble for Lovelock. Never mind that the education of his four children was disrupted again. Never mind that he became more indebted than at any point in his life. He and Helen had to flee.

As a dutiful son, he still divided his time so he could keep Nellie company, but this created new strains. He would spend weekends in the countryside with his family and stay at his mother's flat while working in Mill Hill during the week. In between, he would drive furiously for 100 miles to London every Monday morning and then dash back to Bowerchalke on Thursday night. This could not last. He started smoking and drinking more heavily. As the dowdy 1950s gave way to the free-spirited 1960s, he yearned for a change.

There had never been a stronger conviction in the world that science could enhance and expand the realm of human potential. The power of ingenuity seemed unlimited, no longer even bound by Earthly constraints. Lovelock wanted to be part of this moment in history. And with his matchless ability to decipher the chemical and biological codes of air, water and soil, he was perfectly equipped to explore new ways of thinking about life on Earth, and far beyond.

Dian

'Letter from a beloved.'

The cool flame that flickered into Lovelock's life in the summer of 1965 would illuminate his thinking for decades to come. He had no previous inkling of its existence, nor any capacity to imagine its light. But once guided to wonder in its direction, there it was, in his mind's eye: an atmospheric glow suffused with life, burning for billions of years, enveloping the Earth. This planetary flame drew him in and became an object of fascination, devotion and veneration. Of course, it could also burn, but that was part of the attraction. For in that life-giving radiance, he had found an object of desire.

From this emerged the Gaia theory and an obsession with the atmosphere and its relationship with life on Earth. But he could not have seen it alone. Lovelock was guided by a love affair – a cerebral and physical coupling with an American philosopher and systems analyst, Dian Hitchcock who he met at NASA's Jet Propulsion Laboratory (JPL) in California. Like most brilliant women in the male-dominated world of science in the 1960s, Hitchcock struggled to have her ideas heard, let alone acknowledged. But Lovelock listened. And, as he later acknowledged, without Hitchcock, the world's understanding of itself may well have been very different.

Lovelock had arrived at JPL in 1961 at the invitation of Abe Silverstein, the director of Space Flight Programs at NASA, who wanted an expert in chromatography to measure the chemical composition of the soil and air on other planets. For the science-fiction junkie Lovelock, 'It was like a letter from a beloved. I was as excited and euphoric as if at the peak of passion.' He had been given a front-row seat to the reinvention of the modern world.

California felt like the future. Hollywood was in its pomp, Disneyland had opened six years earlier, Venice Beach was about to become a cradle of youth culture, and Bell Labs, Fairchild and Hewlett-Packard were pioneering the computer chip technology that was to lead to the creation of Silicon Valley. JPL led the fields of space exploration, robotics and rocket technology. Predating NASA by more than two decades, it had passed much of its early history under military control as a centre for missile development. In the 1950s, Wernher von Braun, the German scientist who designed the V-2 rockets that devastated London in the Second World War, made JPL the base for America's first successful satellite programme and it was his technology that the White House was relying on to provide the thrust for missions to the moon, Mars and Venus. By 1961, the Sao Gabriel hillside headquarters of JPL had become a meeting place for many of the planet's finest minds, drawing in Nobel winners, such as Joshua Lederberg, and emerging 'pop-scientists' like Carl Sagan.

There was no more thrilling time to be in the space business. Over the previous five years, the world had watched rapt as the United States and Soviet Union competed for extra-terrestrial and technological supremacy. The military applications of the missile program were terrifying, while the possibility of life on other planets stirred excited curiosity. Front-page headlines and big government budgets were guaranteed year after year.

Lovelock had a relatively minor role as a technical adviser but he was the first Englishman to join the most high-profile, and most lucratively funded, of Cold War frontlines. Everyone on Earth had a stake in the USA–USSR rivalry, but most people felt distant and powerless. Three years earlier, Lovelock had listened on his homemade short-wave radio in Finchley to the beep, beep, beep transmission of the USSR's Sputnik, the first satellite that humanity had put into orbit. Now he was playing with the super powers.

The pressure was on. In 1961, the United States was losing the space race. After Sputnik, the Soviet Union had sent a dog – the Moscow mongrel, Laika – into orbit, propelled robots to the moon, and made Yuri Gagarin the first man in space. Such was the concern in Washington of coming second that the new President John Kennedy opened the cash spigot for space exploration in a way never seen before or since. Congress approved a 25 per cent budget increase for NASA to overtake the Soviets with a US lunar landing and search for life on Mars. Between 1960 and 1966, NASA's share of the US government budget was higher than that for education and welfare. Space was now a national priority. 'I was bound to be carried away,' Lovelock recalled. 'Every few minutes I had to pinch myself to make sure I was there with these other ordinary humans discussing such an extraordinary project. Even a few years previously it would have been unimaginable.'

Lovelock had arrived by helicopter to avoid the Los Angeles traffic. He was put up at the Huntington Sheraton Hotel, a stately edifice located in extensive grounds that was a throwback to the Gilded Age. The trim lawns and manicured hedgerows stood out as an oasis in the deserts of Southern California. Lovelock would spend weeks there at a time, commuting from his hotel room to JPL headquarters on the steep slopes above the parched Arroyo Seco canyon. Built more than twenty years earlier, the cluster of offices and labs had a somewhat tatty appearance, but this was the promised land for scientists. As one JPL recruitment advertisement put it: 'When you were a kid, science fiction gave you a sense of wonder. Now you feel the same just by going to work.'

In the early 1960s, alien life seemed thrillingly possible. Hollywood was churning out films about Martian invasions, and scientists claimed to have identified canals on the red planet that suggested the presence of an extraterrestrial civilisation. The influence of the Victorian-era US astronomer Percival Lowell was still strong, as were his assumptions about the presence of water and moderate atmospheric pressure on Mars. Nobody at JPL expected humanoids or little green men, but there were hopes of finding lower life forms, such as bacteria or microbes.

Lovelock had taken a gamble to be there. The income of a part-time JPL consultant was not enough to support a wife and four children, so he had taken a supplementary job at the University of Houston, Texas. The stipend of $20,000 per year was enough for a spacious house in the suburbs but once again the family had been obliged to move across the Atlantic, with all the disruption that entailed for education and friends. While Jim was roaming the country on lecture tours and fact-finding missions for JPL, Helen was isolated. After two unhappy years, the family decided to return to the UK, leaving Lovelock with an intercontinental commute every few months from his home in the rural village of Bowerchalke, Wiltshire, to California. He flew first class and often landed with a hangover to accompany his jet lag.

Lovelock was, as Nellie grumbled in her diary, little more than a technical adviser on what was to become the Voyager programme, but he was excited by the challenges. The team had to design equipment for an experiment that would be conducted 40 million miles away from the experimenter. It needed to be sufficiently light and small to squeeze inside a 12-cubic-metre capsule along with a 70-watt power supply, communication devices, navigation equipment and two other experiments. It had to be strong enough to withstand forces three times that of gravity during take-off and then, eleven months later, a shuddering landing using Martian atmospheric friction, a parachute and retro rockets to reduce the speed of the capsule from 12,000mph to a bumpy 30mph impact with the surface. The data from the test also needed to be simple because of communication constraints imposed by the radio link.

The head of the JPL bioscience division, Norman Horowitz, said Lovelock's gas chromatograph mass spectrometer was probably the most important single instrument on the Mars lander.

Within a year, Lovelock felt confident enough to send a letter to his boss in which he inveighed against the heavy, unreliable equipment JPL was using. Horowitz asked him to develop an upgrade, a lightweight device that could pump gas out of the mass spectrometer. At his home laboratory in Bowerchalke, Lovelock tested a variety of materials and designs, eventually trying a palladium coil to suck the hydrogen out of the device. He tested the gas that was being emitted by putting it to a flame and hearing the familiar pop of hydrogen. Then as he increased the electric charge to 200 volts, the gas bubble stopped swelling and, at 220, to his amazement, the tiny device started sucking hydrogen up into the tube. At the time he was not quite sure why it worked, but happily put it down as a triumph of intuition. 'The gas vanished into space. I think the hydrogen atoms became protons and floated through the palladium like an electric current. It's magic, really. I now think it was a quantum effect. They may be more ubiquitous than we realise.' He flew back to America to test the new device in front of a crowd of technicians. 'It worked like a dream,' he recalled. 'I got a cheer from the other engineers. Horowitz was very pleased. It is why they were able to do their Mars experiments. I was delighted.'

He celebrated that night in Los Angeles with a new friend, Dian Hitchcock. She was an undercover inspector, who had been hired by NASA to scrutinise the life-detection work being done at JPL. The two organisations had been at loggerheads since 1958, when JPL had been placed under the jurisdiction of the then newly created civilian space agency, NASA, and day-to-day management by the California Institute of Technology. JPL's veteran scientists bristled at being told what to do by their counterparts in the younger, but more powerful federal organisation. NASA was determined to regain control. Hitchcock was both their spy and their battering ram. Lovelock became her besotted ally.

Jim and Dian had met in the JPL canteen, where Hitchcock introduced herself to Lovelock with a joke: 'Do you realise your surname is a polite version of mine?' The crude but witty question delighted Lovelock. As they got to know one another on this and subsequent occasions, he also came to respect Dian's toughness in her dealings with her boss, her colleagues and the scientists. He later saw her yell furiously at a colleague in the street. 'They were frightened of her. NASA were very wise to send her down,' he recalled. They found much in common. Both had struggled to find intellectual peers throughout their lives. Both considered themselves introverted and had encountered learning problems.

Hitchcock, who was eleven years younger, was born Dian Ellen Roark in Long Beach, California, in 1930, and raised during the Great Depression. Her father was a chemical engineer with Texaco oil company and her mother a schoolteacher. Her elder sister had behavioural problems and bouts of rage. Dian had a form of dyslexia but excelled at school. Up until the age of nine, her childhood was happy. Then her mother was put in a mental asylum. 'All she needed was a course of hormones, but they gave her far too many electric shocks,' Hitchcock remembered.

Her father sent Dian, then still a young child, to his sister Jospehine's home in Douglas, Arizona. The young girl felt isolated, miserable and vulnerable. She was forbidden to receive letters from her mother, or even to mention her name. The situation took a sinister turn after she reached puberty, when her uncle started to creep into her room at night: 'I would wake up and see him lifting the covers to look at my body. That scared the shit out of me, but they wouldn't let me lock the door.'

An opportunity to escape arrived in dramatic fashion at the age of fourteen, when her mother was discharged from the asylum. 'I heard her step on the wooden porch and I knew immediately it was her. I called out: "Mother!" and she said: "Let's go! Now!" Three years later they were living in Temple City, California, where she attended the local high school, edited the school paper and, by her own account, came top in an IQ test given to all junior high and high school students in Los Angeles County: 'My score was the extreme outlier, more than six standard errors above the mean.'

As one of the brightest academic lights of her generation, she ought to have had the pick of employers, but as a woman, her career prospects were blighted by chauvinism and worse. She was accepted into Stanford University, where she studied philosophy, earned a scholarship and graduated in 1953. Her final week was traumatic. She was reluctant to divulge what happened: 'Women who stand out with intellectual gifts are punished. I don't want to say too much. It was too scary.'

After a series of short-term jobs, she took up a post at Howard Hughes's aircraft company and, in 1955, married a Berkeley grad student, Tom Hitchcock, who she supported through college by helping to develop a magnetic drum computer at the Marchant Calculator Company. The couple divorced five years later, by which time she had already quit Marchant and returned to Los Angeles because her boss refused to acknowledge her contribution.

There was no shortage of other openings. Hitchcock's natural astuteness and philosophical training gave her an edge in the then fashionable fields of management science and systems design. She joined the Thompson Ramo Wooldridge corporation, which developed computers for the military, and she frequently visited the Pentagon for confidential briefings and consultations. It was during one of these trips to Washington that she met Orr Reynolds, a former US secret intelligence officer and head of the Bioscience Programs Division at NASA.

Orr wanted to tap her formidable intellect. He introduced her to John Lilly, the controversial psychoanalyst who asked her to draw up a programme for initiating conversation with dolphins.

Although her ideas helped secure five years of state funding, Hitchcock's role once again went unrecognised.

Orr hired her as a Hamilton Standard consultant to assess whether JPL's approach to finding life on Mars was based on sound principles or wishful thinking. 'He wanted me to go to JPL and spy on them, investigate everything I could, and report back to him,' she recalled. She took her cat, Blackstone, and rented a big house in Pasadena with the family of her 'nominal boss', Gordon Thomas. Living together quickly proved unbearable. She had been squeezed into the maid's quarters and escaped as often as she could, flitting instead between the homes of other friends until she made a base at the Huntington Sheraton Hotel, although at different times from Lovelock.

She was scornful of the Mars experiments being proposed by the biologists at JPL, which simply replicated what they had been doing on Earth. Where, she asked, was the imagination to consider how different life must be on another planet? Where were the controls that could prevent unreliable or vague results? Where was the big picture that could categorically determine whether Martians existed? Although she did most of the work, she reported to Thomas. When bonuses were handed out at the end of the year, she was overlooked. When she protested, she was told her share had gone to her partner because he needed to support a family and she did not. 'That was how life was then,' she lamented.

Hitchcock had grown used to being overlooked or ignored. In what was then very much a man's world of space exploration, she struggled to find anyone who would take her seriously. That and her inability to find people she could talk to on the same intellectual level left her feeling lonely. Lovelock seemed different. He came across as something of an outsider and was more attentive than other men. 'I was initially invisible. I couldn't find people who would listen to me. But Jim did want to talk to me and I ate it up,' she said. 'When I find someone I can talk to in depth it's a wonderful experience. It happens rarely.'

They were both in their prime. Lovelock was a short, dapper, bespectacled forty-five-year-old at the height of his intellectual and physical powers, who spent his spare time hiking up hillsides. Hitchcock was a trim thirty-four-year-old with sun-bleached hair that she wore in a French twist, whose petite frame belied a forceful presence that was capable of intimidating her superiors. She drew attention for another reason: 'I was well known for huge boobs. I got those from my Irish background.'

Decades later Lovelock claimed he could not even remember the colour of Hitchcock's hair. What attracted him, he said, was not her looks, but her wit and erudition. 'It was her speech that attracted me. I would go as far as to say: "At last I've found someone I can talk to,"' Lovelock recalled. 'She was oceans more intelligent than anyone I'd ever met. She could pull me up. Not many could . . . Conversation with her was a delight. Her control of English was extraordinary. She had all the qualities I wanted in a woman and she fell for me as I did for her.'

They became not just collaborators but conspirators. Hitchcock's scepticism about JPL's approach to finding life on Mars chimed with Lovelock's complaints about the inadequacy of the equipment. This set them against powerful interests. At JPL, the most optimistic scientists were those with the biggest stake in the research. Vance Oyama, an effusively cheerful biochemist who had joined the JPL programme from the University of Houston the same year as Lovelock, put the prospects of life on Mars at 50 per cent. He had a multi-million-dollar reason to be enthusiastic as he was responsible for designing one of the three life-detection experiments on the Mars lander: a small box containing water and a 'chicken soup' of nutrients that were to be poured onto Martian soil. Also upbeat was his colleague Wolf Vishniac, who designed another elaborate test – the 'Wolf Trap' – to collect and grow any micro-organisms that might be found in Martian dirt.

The puppyishly enthusiastic astronomer Carl Sagan said the mission was driven by humanity's need to prove we are not alone in the universe: 'Human beings love to be alive and we have an emotional resonance with something else alive, rather than a molybdenum atom,' he said. 'Why do we go to Antarctica to find out what the emperor penguins have been doing lately? It's fun, because we are primarily drawn to things that are alive.'[1] The wave of optimism hit a peak during a 1964 Summer Study convened by NASA at Stanford University, where for the first time exobiology (the study of life outside of Earth) was recognised as a serious area of academic study. Senior NASA manager Homer Newell declared 'the search for extraterrestrial life' to be a primary goal of the space agency and 'the most exciting, challenging, and profound issue not only of the century but of the whole naturalistic movement'.[2]

To inject a dose of realism into this debate and challenge the Earth-centric scientific approach of JPL, Hitchcock needed an ally. Lovelock fit the bill perfectly. As a technician, he was well liked and respected, but helpfully not embedded into any of the theoreticians' cliques; as a free thinker, he was open to philosophical challenges; and, as a scientist who spent most of his life on the other side of the Atlantic, he had less at stake in the political tussles between NASA and JPL. Hitchcock felt she could use him: 'When I dealt with Jim, I was always intellectually dominant . . . It was always my ideas we talked about.'

In the summer of 1964 or 1965,[3] she wrote to Lovelock, confiding her frustration with JPL's approach and encouraging him to arrange a symposium in England of scientists willing to discuss problems of geocentricity and consider alternative ways to search for life on Mars using information theory. 'I said it was mostly your idea,' she wrote. 'We must recognise, however, that if we have such a meeting and do not clutter it up with the American vested interests we may arouse a hornets' nest . . . you of course will be the primary target.' She closed with an affectionate postscript: 'The JPL study hasn't been nearly as interesting since you left,' which indicates how much the two conspirators were enjoying each other's company by then.

The symposium never went ahead. Hitchcock's boss, Gordon Thomas, was less than impressed by Lovelock's suggestion that the event be held in his village pub, The Bell Inn in Bowerchalke. Lovelock was warned by colleagues that it would be 'very rash' to go ahead with a sceptics' symposium.

Even without a conference, there was growing evidence that the red planet was lifeless. On 15 July 1965, Mariner 4 became the first spacecraft to fly by Mars and return images and data, which revealed a cratered, lunar-like surface and atmospheric pressure that was just 0.4 per cent of that of the Earth, which meant there was little chance of any liquid water. *U.S. News & World Report* proclaimed: 'Mars is dead.'[4] A few months later, infra-red telescopic charts from the Pic du Midi Observatory in France showed the chemical composition of the Martian air was almost entirely carbon dioxide. There was no sign of the lively exchange of gases that signifies the vitality of the Earth. Mars was in seemingly lifeless equilibrium. This was a huge blow to the life-experiment teams, whose equipment now seemed an extravagant folly. Budgets were trimmed and the schedule for Voyager's Mars landing was delayed to 1973. For Lovelock and Hitchcock, this was good news. With their caution now vindicated, their influence and confidence grew.

Dian's relationship with Jim entered a new phase. 'He was very special to me and I was very much in love with him. We had fun. He can be very loving,' Hitchcock reminisced. 'During much of 1965 Jim would fly to the States and then come to Hartford and stay with me for a few days. Then he would fly on to California, and I would arrive a few days later. I remember the staff there would be enthusiastic and tell me about "Jim's latest ideas" which, of course, were the ones we had worked on together. I don't recall ever being even slightly irritated by this. Sharing and producing ideas with Jim was a kind of magical experience, enormously rewarding just on its own.'

At Hitchcock's prompting, her employer, the NASA contractor Hamilton Standard, hired Lovelock as a consultant, which meant she wrote the cheques for all his flights, hotel bills and other expenses during trips to JPL. His role was now far more than a technician. As his Mill Hill laboratory assistant Peter Simmonds put it, he was now 'among the suits'. That meant more responsibility and longer hours when he visited Dian at her home. Lovelock felt he had to up his game: 'I had never before worked from immediately after breakfast, right through until nearly midnight, but the pressure was on.'

On 31 March 1965, Hitchcock submitted a scathing initial report to Hamilton Standard and its client NASA, describing the plans of JPL's bioscience division as excessively costly and unlikely to yield useful data. She accused the biologists of 'geocentrism' in their assumption that experiments to find life on Earth would be equally applicable to other planets, and she suggested a top-down alternative. She felt that information about the presence of life could be found in signs of order – in homeostasis – not in one specific surface location, but at a wider level. As an example of how this might be achieved, she spoke highly of a method of atmospheric gas sampling that she had 'initiated' with Lovelock to detect signs of entropy reduction. 'I thought it obvious that the best experiment to begin with was composition of the atmosphere,' she recalled. This plan was brilliantly simple and thus a clear threat to the complicated, multi-million-dollar experiments that had been on the table up to that point.

At a JPL strategy meeting, Lovelock weighed into the debate with a series of withering comments about using equipment developed in the Mojave Desert to find life on Mars. 'One does not seek fish in a desert or a cactus on an ice-cap,' he said in evident mockery of the Wolf Trap and Oyama's 'chicken soup' experiments. Instead, he proposed the alternative method he had discussed with Hitchcock. In other words, an analysis of planetary gases to assess whether the planet was in equilibrium (lifelessly flatlining) or disequilibrium (vivaciously erratic) based on the assumption that life discharged waste (excess heat and gases) out into space in order to maintain a habitable environment. This led to tensions between Lovelock and the biologists. 'He knew they were full of rubbish,' recalled Peter Simmonds, who had now been brought in to JPL by Lovelock. 'He told them the key was disequilibrium and said: "Don't scrabble in the soil, focus on atmospheric analysis." He made himself very unpopular.'

The critique caught the attention of a technical director, Robert Meghreblian. He asked Lovelock what he would do differently and, crucially, what equipment he would use to do it. Jim said he responded within days with a tiny device that would heat up the surface of Mars and analyse whether the atmosphere was capable of supporting life. It was a gas chromatograph analysis of entropy reduction. This would be the basis for a submission to *Nature*, which appeared on 1 August 1965 under the title 'A Physical Basis for Life Detection Experiments'. This was solely in his name, though it echoed many of the ideas that Dian had elaborated in her earlier internal report. It was Lovelock's first publication as an independent scientist and he later described it as 'the first Gaia paper', though the embryonic theory did not yet have a name. The recognition by this prestigious journal boosted his status and gained support for his and Hitchcock's proposal of a 'top-down' approach to looking for signs of life.

Dian later complained that she deserved more credit, but she said nothing at the time. These were giddy times for the couple. They were spending more days together and were clearly fascinated by one another. Hitchcock took Lovelock to a lakeside farm owned by Warren McCullock, the idiosyncratic neurologist and founder of the American Society for Cybernetics. 'She got interested in whether I have an eidetic memory,' Lovelock recalled. 'We had a lovely time. They examined my mental health and brain function . . . That illustrates the kind of environment she worked in.' Hitchcock's only recollection of that weekend is rather different: 'I do recall that Warren didn't approve of my relationship, and warned me that I might end up regretting it. As I was deeply in love with Jim at the time, I paid no attention.'

The melding of intellects was now also a sharing of bodies. 'Our trysts were all in hotels in the US,' Lovelock remembered. 'It was a good sex relationship. We carried on the affair for six months or more.' Romance and science were interwoven. Their pillow talk was imagining how a Martian scientist might find clues from the Earth's atmosphere that our planet was full of life. This was essential for the Gaia hypothesis. Hitchcock had posed the key question: what made life possible here and, apparently, nowhere else? This set them thinking about the Earth as a self-regulating system in which the atmosphere was a product of life. Instead of the wonder of boundless space, the couple dwelled on the astonishing limits that had made life possible on Earth but nowhere else.

From this revolutionary perspective, the gases surrounding the Earth suddenly began to take on an air of vitality. They were not just life-enabling, they were suffused with life, like the exhalation of a planetary being – or what they called in their private correspondence, the 'great animal'. Far more complex and irregular than the atmosphere of a dead planet like Mars, these gases burned with life. This was not just a metaphor. Lovelock later discovered that in the upper strata of the atmosphere, ultraviolet sunlight could, in effect, ignite methane and oxygen, creating a 'cool flame' that had been burning for billions of years.

They sounded out others. Sagan, who shared an office with Lovelock, provided a new temperature-regulating dimension to their idea by asking how the Earth had remained relatively cool even though the sun had steadily grown hotter over the previous 8 billion years. Lewis Kaplan at JPL and Peter Fellgett at Reading University were important early allies and listeners, as, in his own way, was Horowitz. The long-dead physicist Erwin Schrödinger also provided an important key, according to Lovelock: 'I knew nothing about finding life or what life was. The first thing I read was Schrödinger's *What is Life?* He said life chucked out high entropy systems into the environment.' That was the basis of Gaia; I realised planet Earth excretes heat.'

In the mid 1960s, this was all still too new and unformed to be described as a hypothesis. But it was a whole new way of thinking about life on Earth. They were going further than Charles Darwin in arguing that life does not just adapt to the environment, it also shapes it. This meant evolution was far more of a two-way relationship than mainstream science had previously acknowledged. Life was no longer just a passive object of change; it was an agent. The couple were thrilled. They were pioneers making an intellectual journey nobody had made before. It was to be the high point in their relationship.

The following two years were a bumpy return to earth. Lovelock was uncomfortable with the management duties he had been given. The budget was an unwelcome responsibility for a man who struggled with numbers, and he was worried he lacked the street smarts to sniff out the charlatans who were pitching bogus multi-million-dollar projects. Meanwhile, the biologists Oyama and Lederberg were going above his head and taking every opportunity to put him down. 'Oyama would come up and say: "What are you doing there? You are wasting your time, NASA's time,"' Lovelock recalled. 'He was one of the few unbearable persons I have known in my life.' In 1966, they had their way and Lovelock and Hitchcock's plans for an alternative Mars life-exploration operation using atmospheric analysis by infra-red telescopes were dropped by the US space agency. 'I am sorry to hear that politics has interfered with your chances of a subcontract from NASA,' Lovelock's colleague Peter Fellgett commiserated.[5]

Cracks started to appear in Jim's relationship with Dian. He had tried to keep the affair secret, but marital lies weighed heavy: 'I'm not suited for double relationships. I'm not good at cheating.' They could never go to the theatre, concerts or parks in case they were spotted together, but close friends could see what was happening. 'They naturally gravitated towards one another. It was obvious,' Simmonds said. When they corresponded, Lovelock insisted Hitchcock never discuss anything but work and science in her letters, which he knew would be opened by his wife, who worked as his secretary. But intimacy and passion still came across in discussions of their theories.

Lovelock's family noticed a change in his behaviour. The previous year, Nellie suspected her son was unhappy in his marriage and struggling with a big decision. Helen openly ridiculed his newly acquired philosophical pretensions and way of talking – both no doubt influenced by Hitchcock. 'Who does he think he is? A second Einstein?' she asked scornfully. Helen was suspicious about his relationship with Dian, but Jim denied everything.

On at least one occasion he invited Dian to the family home in Bowerchalke during a work trip to the UK. When they went out for a drive together, Dian insisted on sitting in the front next to Jim, relegating his wife to the backseat. 'It was just what Grandma used to do,' Christine sighed.

Helen sensed trouble. She would scornfully refer to Hitchcock as 'Madam' or 'Fanny by Gaslight', forbade her husband from introducing Hitchcock to other acquaintances, and insisted he spend less time in America. But he could not stay away, and Helen could not help but fret: 'Why do you keep asking me what I'm worried about? You know I don't like (you) all those miles away. I'm only human, dear, and nervous. I can only sincerely hope by now you have been to JPL and found that you do not have to stay anything like a month. I had a night of nightmares . . . The bed is awfully big and cold without you.' A couple of weeks later, she took aim at everyone and everything keeping her husband away from her: 'Bugger Ab [Zlatkis – Lovelock's business partner], bugger Houston, bugger JPL and bugger money!' By 25 May, Helen was barely able to contain her misery. 'What a dreadful life you have married to me,' she wrote. 'The days are up – WHEN???' She forbade him from visiting the US for more than five days.[6]

So, Lovelock visited JPL less frequently and for shorter periods. Hitchcock filled the physical void by throwing her energy into their shared intellectual work. Taking the lead as usual, she began drafting a summary of their life-detection ideas for an ambitious series of journal papers about exobiology[7] that she hoped would persuade either the US Congress or the British parliament to fund a 100-inch infra-red telescope to search planetary atmospheres for evidence of life.

Nothing seemed to be going their way. In successive weeks, their jointly authored paper on life detection was rejected by both the *Proceedings of the Royal Society* in the UK and then *Science* in the US. The partners agreed to swallow their pride and submit their work to the little-known journal *Icarus*. Hitchcock admitted to feeling downhearted in a handwritten note:

11 November 1966

Enclosed is a copy of our masterpiece, now doubly blessed since it has been rejected by *Science*. No explanation so I suppose it got turned down by all the reviewers . . . Feel rather badly about the rejection. Have you ever had trouble like this, publishing anything? . . . As for going for *Icarus*, I can't find anybody who's even heard of the journal.

Hitchcock refused to give up. Maybe she wanted to remind Jim of the mental luminosity that had first drawn them together. Maybe she wanted to show how much she still cared. Or perhaps she wanted their relationship to leave something behind. In late 1966 and early 1967, she sent a flurry of long, intellectually vivacious letters to Lovelock about the papers they were working on together. Her correspondence during this period was obsessive, hesitant, acerbic, considerate, critical, encouraging and among the most brilliant in the Lovelock archives. These missives can be read either as foundation stones for the Gaia hypothesis or as thinly disguised love letters.

In one she lamented that they were unable to meet in person to discuss their work, but she enthused about how far their intellectual journey had taken them. 'I'm getting rather impressed with us as I read "Biology and the Exploration of Mars" – with the fantastic importance of the topic. Wow, if this works and we do find life on Mars we will be in the limelight,' she wrote. Further on, she portrayed the two of them as explorers, whose advanced ideas put them up against the world, or at least against the senior members of the JPL biology team. 'We are the only people who are talking about planetary exploration and the role of MIFS [the Multiplex Interferometric Spectroscopy that was the basis of their proposed telescope project].'

The most impressive of these letters is a screed in which Dian wrote to Jim with an eloquent summary of 'our reasoning' and how this shared approach went beyond mainstream science. 'We want to see whether a biota exists – not whether single animals exist,' she said. 'It is also the nature of single species to affect their living and non-living environments – to leave traces of themselves and their activity everywhere. Therefore we conclude that the biota must leave its characteristic signature on the 'non-living' portions of the environment.' Hitchcock then went on to describe how the couple had tried to identify life:

13 December 1966

We started our search for the unmistakable physical signature of the terrestrial biota, believing that if we found it, it would – like all other effects of biological entities – be recognisable as such by virtue of the fact that it represents 'information' in the pure and simple sense of a state of affairs which is enormously improbable on non-biological grounds ... We picked the atmosphere as the most likely residence of the signature, on the grounds that the chemical interactions with atmospheres are probably characteristic of all biotas. We then tried to find something in our atmosphere which would, for example, tell a good Martian chemist that life exists here. We made false starts because we foolishly looked for one give-away component. There are none. Came the dawn and we saw that the total atmospheric mixture is a peculiar one which is in fact so information-full that it is improbable. And so forth. And now we tend to view the atmosphere almost as something itself alive, because it is the product of the biota and an essential channel by which elements of the great living animal communicate – it is indeed the milieu internal which is maintained by the biota as a whole for the well-being of its components. This is getting too long. Hope it helps. Will write again soon.

With hindsight, these words are astonishingly prescient and poignant. Their view of the atmosphere 'almost as something itself alive' was to become a pillar of Gaia theory. The connection between life and the atmosphere, which was only intuited here, would be firmly established by climatologists. It was not just the persuasiveness of the science that resonates in this letter, but the intellectual passion with which ideas are developed and given lyrical expression. 'We' is the most oft-repeated word on the page, as if their togetherness is a concept that the author wanted to hammer home again and again, almost as forcefully as the life-detection theory. The poetic conclusion – 'came the dawn' – reads as both a hopeful burst of illumination and a sad intimation that the night together, of 'we'-ness, may be drawing to a close.

Their joint paper, 'Life detection by atmospheric analysis', was submitted to *Icarus* in December 1966. Lovelock acknowledged it was superior to his earlier piece for *Nature*: 'Anybody who was competent would see the difference, how the ideas had been cleared up and presented in a much more logical way.' He insisted Hitchock be lead author: 'Dian was a great writer, very fluent and rapid. She took over that side of the work. I was adamant that my name shouldn't come first.' She concurred. Although glad to have him on board because she had never before written a scientific paper and would have struggled to get the piece published if she had put it solely under her name, she had no doubt she deserved most of the credit: 'I remember when I wrote that paper, I hardly let him put a word in. I said: "Jim you are not getting the point of this." I eliminated all of his writing. I rewrote a draft and rewrote it and rewrote it.' She said she also over-rode Lovelock by insisting that they cite Evelyn Hutchinson, the British scientist often described as the 'father of modern ecology', as the first person to point out that the combination of oxygen and methane in Earth's atmosphere is evidence of life.

The year 1967 was to prove horrendous for them both, professionally and personally. In fact, it was a dire moment for the entire US science programme. In January, three astronauts died in a flash fire during a test on an Apollo 204 spacecraft, prompting soul-searching and internal investigations. US politicians were no longer willing to write blank cheques for a race to Mars. Public priorities were shifting as the Vietnam War and the Civil Rights movement gained ground. Congress slashed the NASA budget by $500 million. It was a final blow for the Voyager programme, which was to die a bureaucratic death in an appropriations committee the following year.

The affair between Hitchcock and Lovelock was approaching an ugly end. They had never professed their love for one another because both knew it could not last. Domestic pressures were becoming intense. Helen was increasingly prone to illness and resentment. On 15 March 1967, she wrote to Jim at JPL to say: 'It seems as if you have been gone for ages,' and scornfully asked about Dian: 'Has Madam arrived yet?' Around this time, Lovelock's colleague at JPL, Peter Simmonds, remembered things coming to a head. 'He strayed from the fold. Helen told him to "get on a plane or you won't have a marriage" or some such ultimatum.'

Lovelock was forced into an agonising decision about Dian. 'We were in love with each other. It was very difficult. I think that was one of the worst times in my life. Helen was getting much worse. She needed me. It was clear where duty led me and I had four kids. Had Helen been fit and well, despite the size of the family, it would have been easier to go off.' Instead, he decided to ditch Dian. 'I determined to break it off. It made me very miserable . . . I just couldn't continue.'

Hitchcock had other worries. Her plans to visit Lovelock in England in the spring of 1967 had to be put on hold when she discovered a lump on her left breast which required surgery. Ominously, Jim had become impossible to get hold of. On 25 April, she wrote: 'Tried to call you this AM but no luck as you are in Holland, it seems . . . [My] trip postponed 1 June to about 16 June . . . let me know if and when you'll have free time during this period.' There is no record of a reply.

The break-up, when it finally came, was brutal. Today, more than fifty years on, Hitchcock is still pained by the way things ended. 'I think it was 1967. We were both checking into the Huntington and got rooms that were separated by a conference room. Just after I opened the door, a door on the opposite side was opened by Jim. We looked at each other and I said something like: "Look, Jim, this is really handy." Whereupon he closed the door and never spoke to me again. I was shattered. Probably "heartbroken" is the appropriate term here. He didn't give me any explanation. He didn't say anything about Helen. He just dropped me. I was puzzled and deeply hurt. It had to end, but he could have said something . . . He could not possibly have been more miserable than I was.'

Hitchcock was reluctant to let go. That summer, she sent Lovelock a clipping of her interview with the local Connecticut newspaper,[8] with a handwritten note: 'Full of errors and a bad photo but at least it's space.' Below the headline, 'A Telescopic Look at Life on Other Planets', is a photo captioned 'Space explorer Mrs Diana (sic) R. Hitchcock' and an article outlining the bid she and Lovelock were preparing in order to secure financial support from Washington for a telescope. In November, she wrote a memo for her company detailing the importance of her continued collaboration with Lovelock and stressing their work 'must be published'.

But the flame had been extinguished. The last record of direct correspondence between the couple is an official invoice, dated 18 March 1968, and formally signed 'consultant James E Lovelock'. Hitchcock was fired soon after by Hamilton Standard. 'They were not pleased that I had anything at all to do with Mars,' she recalled. The same was probably also true for her relationship with Jim.

Lovelock had other things on his mind. Helen was increasingly unwell but now insisted on joining her husband on work trips. Two years later, she was with him at a hotel in Chester when she experienced a sudden, frightening loss of vision and balance. At the insistence of Jane, who was now a nurse, the family doctor arranged a lumbar puncture in Southampton. Lovelock drove her to the hospital. According to his daughters, he was once again in a foul mood because he was worried and felt powerless. A week later, the long mystery of Helen's illness was revealed as multiple sclerosis. She was relieved to finally have a diagnosis, even if it was life-changing and life-shortening. Lovelock drew up a project to maintain his wife's health. To keep the symptoms at bay, she changed her diet and started an exercise regime, promising to 'walk ploughed fields and hang by the arms from ladders', according to her daughter Christine.

In the half-century that followed, Jim met Dian just once more – an awkward chance encounter at a conference in Europe. By then they had been proved right about the absence of life on the red planet. After Voyager was abandoned, the Mars programme stuttered on in the form of a scaled-down Viking mission that finally landed in 1976. Lovelock had already left NASA, but one of his gas chromatographs made it onto the surface. The Wolf

Trap was not so lucky. It had been culled in earlier budget cuts. Oyama had held on and now, finally, had the chance to lure Martians with his 'chicken soup'. There were no takers. Not even a micro-sip from a micro-organism. Just red dust. Even Carl Sagan had to admit defeat: 'There was not a hint of life – no bushes, no trees, no cactus, no giraffes, antelopes, or rabbits, no burrows, tracks, footprints, or spoor; no patches of colour that might be photosynthetic pigments.'[9]

NASA scientists began fighting about the 'terribly ad-hoc'[10] interpretation of data and the insensitivity of Oyama's device. It marked the end of the Lederberg school of exobiology. Scientists began to accept that the question of life on Mars was more likely to be resolved, as Hitchcock and Lovelock had suggested, by telescopes on Earth than instruments on the distant planet. The most common takeaway from the disappointing mission was that maybe life on Earth really was unique, after all. As Horowitz later concluded: 'The origin of life may have happened only once, and it happened here and no place else in the solar system. Or if it happened elsewhere, it didn't survive. I think this is a conclusion of really cosmic importance. If people become aware of this, then maybe they'll be less inclined to destroy the planet.'[11]

It was all exactly as Hitchcock and Lovelock had foreseen, but this time there was no chance of them celebrating their foresight together. In the 1970s, Dian remained in the field of atmospheric chemistry, writing a handful of papers and then marrying her second husband, the US paleolimnologist Edward Deevey. He died six years later, after which she became active in local politics and ecology, campaigning against a power plant in her home district of Gainesville, Florida, at the age of eighty.

She remains an intellectual force to be reckoned with and argues that her founding role in the Gaia hypothesis remains unappreciated because she is a woman – and women were rarely taken seriously in the 1960s: 'At the time I was in love and in the 1960s women were supposed to be happy when their men achieved success. But looking back, I think my contribution deserved a credit. Often men would talk to me and say: "I love talking to you. I get so many good ideas," but they were my ideas goddammit . . . Jim couldn't have been immune to that, though he certainly never would have said that at the time. I wouldn't have let him get away with that.' She adds: 'He was a kind man who wouldn't do anything sneaky or dirty or wrong. It does not fit my real memories of him.'

Who conceived which ideas is difficult to determine after so many years. There was a constant to and fro of thoughts in person and by mail, not all of which survive. It is a fact that several of the key philosophical concepts in Lovelock's *Nature* paper, particularly entropy reduction, had already been published in Hitchcock's internal report for Hamilton Standard. But it is also true that Lovelock was more knowledgeable and experienced in chemistry, chromatography and instrument design, which were all crucial to this study.

He also contributed a solitary imagination to the question of life detection. While Sagan and other scientists at JPL were reluctant to accept the Earth might be the only planet with life, Lovelock had no difficulty embracing the possibility of being alone in the solar system. Childhood had taught him that loneliness was a natural state. While the other scientists at JPL remained fixated on looking outward for alien life, he reflected inwards on his home planet and asked himself the question posed by Hitchcock: what made life possible here and, apparently, nowhere else? This set the two of them thinking about the Earth as a self-regulating system in which the atmosphere was a product of life, not just part of an environment to which it had to adapt. For them, outer space exploration had led to greater self-questioning and awareness, to heightened sentience.

Both conceded their memories of this period were imprecise, which is hardly surprising after so many years and the muddying effect of break-ups, social niceties and subsequent relationships. What is clear is that Hitchcock deserves more recognition for her essential role in the conception and early development of Gaia theory. The year after they split up, Lovelock co-authored a paper[12] with the JPL engineer C.E. Giffin that contained many of the brilliant phrases and suggestions Dian had introduced during her correspondence with him. He used them again in a lecture on planetary ecology the following October, when he made the first public link between the 'biological cybernetic system' (an early, clunky name for Gaia) and environmental concerns. He began that speech by looking back at the idea's conception and the partner who helped bring it into existence: 'It all started in 1964 when, with a colleague, Dian Hitchcock, we wondered if it would be possible to detect the presence of life on Mars, if any was there to be detected.'

In 1969, he published a letter in *Scientific American* to 'straighten the record' about Hitchcock's role, stating that 'she is due at least half the credit' for the discovery that the atmosphere of the Earth is an actively controlled component of its ecosystem. But it was only at the end of his life that Lovelock finally opened up about how much Hitchcock had meant to him personally: 'She was one of the first confidantes on what was to be Gaia theory . . . She undoubtedly helped me . . . I was very fond of her. I dearly wish something could have come of it. I couldn't do it. It tore me apart. I didn't confide in a soul. It was bloody hard. I think she helped my mental education more than I really care to say.'

He had never told Hitchcock of his adoration for her and the domestic difficulties that eventually forced them apart. This was passed on by the author of this book. Finally learning of Jim's feelings at the age of ninety-two, Dian was so astonished that she sobbed: 'I am overwhelmed at the idea that Jim loved me. We did have an affair, and I was very much in love with him, but he never suggested in any way that there could ever be some long-term possibilities for us . . . Until you told me that Jim really cared deeply for me, I believed the only person who ever loved me was my mother.'

The doomed romance could not have been more symbolic. Hitchcock and Lovelock had transformed humanity's view of its place in the universe. By revealing the interplay between life and the atmosphere, they had shown how fragile are the conditions for existence on this planet and how unlikely are the prospects for life elsewhere in the solar system. They had brought romantic dreams of endless expansion back down to Earth with a bump.

This raised new questions: How did 'the great animal' of planetary Life (with a capital L) behave? Was there a purpose behind its function? Did it rely on a delicate equilibrium that could be upset by human behaviour or did it possess the power to sustain and regulate itself? The answers to these existential matters might once have been provided by religion, philosophy or science, but by the mid- to late- 1960s business and politics infused everything. The nascent debate about Life coincided with a shift in public consciousness. Science had revealed its own limits. This was a pivotal moment. Just as the race for space now seemed a dubious investment of time and money so the benefits of industry were starting to be weighed against the environmental costs. Lovelock was at the heart of these issues. His sensitive instruments and 'Martian-eyed' perspective of the Earth had given him a unique perspective to consider the future of the planet and the climate risks posed by humanity's disruption of the atmosphere. But his warnings would initially be a commercial secret and his broader vision coloured by another powerful influence on his thinking.

Victor

'So many problems, so few solutions.'

13 April 1967

Dear Lovelock
... I wanted to ask whether you would write a memorandum on what the world will look like in the year 2000 and what effect it will have on a company such as Shell.
Yours sincerely
Rothschild
Telephone: Waterloo 1234

The invitation to be a futurologist was impossible for Lovelock to ignore. By 1967, he was in his intellectual pomp. Although Jim's JPL work was winding down along with his relationship to Hitchcock, he was working concurrently on a host of other projects. Nothing seemed too ambitious, and this proposal was particularly appealing. As a would-be science fiction writer, how could he not be excited by a commission that would give free rein to his end-of-millennium imagination? As one of the leading atmospheric chemists in the world, why wouldn't he want to project forward the disturbing climate trends he had started to observe through his instruments? And as an entrepreneur, how could he say no to his principal client, who also happened to be one of the most powerful men in the world.

Baron Victor Rothschild had hired Lovelock four years earlier as a one-man think tank, a consultant on all and any matters of interest to Shell, the petrochemical multinational in which his family was a major shareholder. The scion of the Rothschild dynasty was a prominent scientist himself and the company's director of research, which put him on the frontline of growing environmental criticism of the oil industry. He was also extraordinarily well connected to the secret services on both sides of the Atlantic and beyond.

Lovelock had previously sent him one of the earliest and most detailed warnings of the destabilising effect of fossil fuels on the global climate. At Rothschild's behest, he had interviewed the leading meteorologists in the US and the UK, gained access to unpublished studies and conducted his own research into atmospheric pollution. His findings were unveiled in three reports that alerted his client to the dangers posed by Shell's business activities. As early as 1966, Lovelock told Rothschild of 'the almost certain fact that the climate is worsening and the probability that the combustion of fuel is responsible'; discussed the 'disturbing' possibility of 'positive feedbacks' in the climate system that could make a bad problem worse; and spelled out that a failure to take action could result in 'serious' hazards and 'unpleasant' consequences.

The forecast for the year 2000 followed on from this and was no more reassuring. Lovelock told the Shell executive that by the end of the century, pollution would be catastrophic, fossil fuels would have to be replaced by nuclear power, and that conservation would become the great political and religious issue of the day.

His millennial 'memorandum' also echoed proto-Gaian phrases from Hitchcock's work with him around this time: 'Until now, the Earth's climate and the chemical composition of its surface, air and sea have evolved with life to provide optimal conditions for its survival, and this optimum has been actively maintained by cybernetic biological processes. It's in Shell's interest to help maintain this optimum, whatever the cause of any deviation from it. It should be an interesting challenge and keep us alive.' These words are intriguing. On the one hand, they read as a not-so-discreet plea for the company to avoid activities that destabilise the optimal conditions of life. On the other, they clearly state it may be in the company's interests to know more about cybernetic biological processes, which would later be called Gaia.

The oil baron told him to keep quiet about his predictions and directed him to a still more ambitious assignment: fixing the global climate.

That proved a challenge too far, even for Lovelock. On deadline day of New Year's Eve 1968, Jim did his best to cobble together suggestions for atmospheric modification, but, in his heart, he knew this geo-engineering pitch was weak, so he added an apologetic note to his submission.

1 January 1969

Dear Lord Rothschild
When I saw you last, in May 1968, I said that I would let you have before the end of the year a note on what might be done to stop the fall in temperature. This was rash of me, for the problem is very complex and long term.

Given what we now know about the impact of fossil fuels on the atmosphere, it may seem strange, even slightly comical, for a scientist to say sorry to an oil baron for failing to solve the climate threat, as if he were a student who had to answer to a teacher for neglecting his homework. It may also sound odd that the primary concern in the late 1960s was the possibility of a new ice age. But the fall in temperatures over the previous decades was the first indication that human activity, especially the burning of oil, gas and coal, was destabilising the climate.

Lovelock had advised Rothschild that fossil-fuel combustion was the cause of atmospheric disruption and other environmental problems, but he did not dare to suggest a phasing-out of petrol sales to Shell. Nor did he immediately raise the alarm among the public. Instead, his warnings were buried for decades alongside other corporate secrets. Asked why he did not act earlier, Lovelock said there was too much uncertainty. 'I wasn't in the habit of warning people about things. We didn't have sufficient measurements. The atmospheric carbon dioxide figure from Mauna Loa [the monitoring station in Hawaii that was used as a global benchmark] was steadily rising but it was still a long way from a level that would cause noticeable climate change. Still, we could predict it might get there.'

That risk nagged at him, but he felt powerless. 'I couldn't put it out of my mind. I knew the climate was changing and what we were doing was having an effect. It wasn't that I wasn't concerned but I didn't think I could do anything.' His feeling of impotence stemmed from an imbalance between knowledge and money, and between science and business. Lovelock felt this most strongly in the client–provider and mentor–protégé relationship with Rothschild, which dominated his working life in the 1960s. The correspondence between the two men, much of it confidential and unpublished until now, provides a blow-by-blow account of how a fossil-fuel company came to realise it was wrecking ecosystems, degrading human health and destabilising the climate, only to cover up the evidence, downplay the risks and blame others.

Their relationship had started in the late summer of 1963 with a mailed invitation that dropped through the Lovelock letterbox in Wiltshire. The contents were terse and somewhat oblique – a style Lovelock would soon become accustomed to: 'Will you come to see me on Thursday at 11 o'clock at Shell Centre on the South Bank to discuss something to your interest?' The message was signed with an indecipherable scrawl, followed by the printed name 'Lord Rothschild'.

A few days later, Jim headed off from Salisbury station for an interview with the sender. As his carriage rattled through the countryside, he knew he was on his way to a life-changing encounter.

The two men had met once, a decade or so earlier, when they were biologists separately studying how to freeze spermatozoa, which was one of Lovelock's areas of research in Mill Hill. Back then, they were scientific equals. This time, Lovelock was very much the social junior seeking patronage. A few months earlier, he had come back from Houston and left the stability of his employer of the previous twenty-two years, the MRC, and returned to Wiltshire. He was now an independent scientist with a wife and four children, living in temporary accommodation and working as a part-time consultant for JPL in the United States. Money was short. He desperately needed a well-heeled client.

Victor Rothschild was one of the world's most powerful fossil-fuel aristocrats. In the late nineteenth century, his Europe-based banking family had established an oil and shipping empire that, at one point, exceeded the global reach of Rockefeller's Standard Oil in the US. It also played an influential role in the founding of Royal Dutch Shell. Victor had inherited the family title and wealth in 1937, but back then he was more interested in sport, science and intrigue than the family's main banking business.

In his youth, he had excelled at everything he turned his hand to. He earned a zoology PhD at Cambridge and played first-class cricket for Northamptonshire until a few too many 'bodyline' blows prompted him to take early retirement from the sport. During the Second World War, he ran espionage rings and disinformation campaigns for MI5 and won a George Medal for dismantling a German booby trap. In the years that followed, he chaired the Agricultural Research Council, where he championed scientific reason over tradition. He was a jazz aficionado, skilled pianist and collector of rare books and manuscripts. In 1961, when he became vice chairman, and then chairman, of Shell Research, he started another collection – this time of very intelligent people.

Victor

As the train pulled into Waterloo station, Lovelock was dazzled by his first sight of the Shell Centre, which had opened the previous year with a design that exuded the growing power of the oil industry. Up until the 1960s, London had retained a largely Renaissance skyline for three centuries. Apart from the dome of St Paul's cathedral and the Victoria Tower of the Houses of Parliament, the only architectural features that came close to scraping the sky before then were church steeples and chimney pots. The Shell Centre changed everything, giving the English capital its first Manhattan-esque tower block and re-ordering the visual hierarchy of the city. At 107 metres, Shell had the first storeyed building in England to rise above the Palace of Westminster on the other side of the Thames. Rothschild, who had the office suite on the top floor, was thus the first business executive who could literally look down on Parliament, Buckingham Palace and the government offices in Whitehall. It was a fitting perspective; the multinational oil corporation was raising an empire as the British one diminished.

From Waterloo station, Lovelock strolled across a custom-built walkway to Shell's reception area. When he told the staff he was there to see Rothschild, he recalled 'everyone snapped to attention' and a 'flunky' quickly escorted him to one of the forty elevators that whisked him up to the twenty-fourth floor where Rothschild's secretary, Miss Page, ushered him into her boss's carpeted office. The former Brixton urchin was awed. 'I was a little bit scared at first. He was the most senior person I'd met,' Lovelock recalled. 'He was an imposing figure. Everyone seems like that to me because I am a shrimp. Rothschild was about 5 foot 10 inches, like my father. He was formidable.'

They sat in armchairs and Rothschild got down to brass tacks. He said he wanted a scientific adviser who he could consult from time to time on matters of interest to Shell. He promised Lovelock could do the work at home, apart from monthly or quarterly meetings with him in London and occasional trips to Shell's research sites at Thornton, near Chester, and Sittingbourne in Kent. In return, he offered an annual consultant's fee of £1,000 (about a third of the average price of a three-bedroom house). Lovelock recalled leaving the office in a 'happy daze' about the new life that suddenly seemed possible after months of uncertainty. 'I'd resigned from a well-paid job with tenure and medical fees. I had four children. It had been madness to think of breaking away at that point . . . But when I went to see Rothschild, it all began to open up.'

Rothschild, who was managing thirty laboratories, had been tipped off that the inventor of the ECD was looking for a new role by Lovelock's former boss, Charles Harrington, who had already helped the Shell executive to recruit two of his colleagues from the NIMR the previous year. It was a sign of the times. The cream of the scientific talent in British government research institutions was pouring into the private sector. Many moved to the United States, where corporations paid three times UK government salaries. This led the British media to bewail a 'brain-drain'. Rothschild expressed patriotic pride that he had kept some of the best talent in the UK. And Lovelock was his greatest prize. In a later letter to a friend, he wrote: 'Sometime, when we meet, I will explain to you how we came to take on Lovelock, who is a "brain-drain" in reverse.'[1]

Their working relationship started in a light mood, but over the years grew heavier along with the weight of their shared secret about the company's effect on human health and the environment. They got together once a month at first, then every quarter. Their meeting of minds was always at Shell Centre, usually in the morning and often stretched into long boozy lunches. They clearly enjoyed each other's intelligence. Rothschild was regarded as one of the brightest men in Britain. During a secondment to the US army in 1944, he had taken an IQ test and later claimed the second-highest score ever registered by the millions of troops who had passed through the ranks. Lovelock's own struggle to find peers dated back to his solitary childhood but now here he was with a fellow polymath. The hierarchy, however, was clear. Rothschild always set the tone and pace of exchanges. In the early years, he dispatched Lovelock to learn the workings of the petroleum industry at Shell laboratories and refineries at Wood River, near St Louis in the US, and the two UK research sites.

Lovelock picked up the mannerisms and beliefs of his mentor, particularly an aversion for committees, which Rothschild described as 'easier than thinking and, therefore, rather seductive', as well as advice on how to avoid tax by 'never having an income'. In the first couple of years, their correspondence is the epitome of brevity: many letters from Rothschild are just one or two lines long, as dry as a Martini, but occasionally spiked with expressions of bathos, self-deprecation or wilful obliqueness. This would not be so strange in a telegram, or later in an email or tweet, but each of these one-liners had to be dictated to a secretary, typed up, put in an envelope, addressed, affixed with a stamp and sent the 95 miles to Bowerchalke. Some of these memos barely merit the role of a nudge:

7 September 1964

Dear Lovelock
What about the muscle machine?

28 December 1964

Dear Lovelock
You have deserted me. Could we not have a talk next time you are in London?

There is a zany, boffin-like air to their early exchanges. Rothschild and Lovelock were exploring potentially patentable inventions that could push back the boundaries of science for commercial gain. This was the new spirit of entrepreneurial technology in the UK, which was belatedly catching up with the innovation trends that had supercharged US economic growth in the 1950s. 'It was lovely working with Victor Rothschild.

We'd pick up lots of crazy ideas,' Lovelock said later. 'In all my time with Rothschild, we never argued. We disagreed only on technical points. I wouldn't argue with a man like that unless I knew I was right . . . We became friends and I changed from Lovelock to Jim.'

Such affinity was uncommon at Shell, where Rothschild could be a drunk and a bully to his underlings. As Lovelock remembered: 'They were all so scared of him that for anyone to be sympathetic to him personally was very unusual . . . He was a very difficult man. He was a misogynist. That was one of his worst points. More than once Miss Page was in tears because she had been in a battle with him. She would sometimes say: "Be careful, he's impossible to deal with today because he has a fierce hangover." I knew he had a bad temper, so I took the initiative. "You look rough. Can I help?"'

In 1965, while Nellie was worrying that her son was throwing his life away as a mere 'technician', Rothschild helped Lovelock find a new client: the British secret service, which was the start of a working relationship with MI5 and MI6 that was to last until well into his nineties and possibly until his 100th birthday. During a visit to the United States at the turn of the year, Lovelock was asked by a colleague at Yale, Sandy Lipsky, whether his ECD could sniff out Viet Cong camps hidden in the jungles of South East Asia. Jim sketched out a system of chemical tracing agents that could be covertly and invisibly applied to hostile agents (either by marking or ingestion) enabling their location to be monitored at a distance by the ECD.

The invention record, which was registered in March 1965 and disclosed in Connecticut to Dian Hitchcock, described how fluorocarbon compounds could be fingerprinted onto the item or person to be traced: 'Where the recipient is an animal – including man – the volatile mixture is released from its encapsulated, dissolved, absorbed or chemically altered form by the normal process of digestion, and will be liberated into the atmosphere during respiration.'[2] The 'scent' of this chemical fingerprint could then be detected by collecting a series of samples of the air at a distance and then located by measuring the wind speed and direction. This prototype system was sensitive enough to detect the evaporation of 1mg of chlorofluorocarbon (CFC) tracer in an area one half-mile square by 160 feet high.

To Lovelock's chagrin, this was not precise enough for the CIA and the Pentagon, who turned down his proposal. However, it intrigued Rothschild, who asked for a memo on the subject, which he passed on to his contacts in the British intelligence community.

Rothschild had long since given up a formal role in MI5, but he had kept his access and influence. As the baron's biographer notes:

> After the war the British secret service, much to the surprise and indignation of the CIA, continued to involve privileged amateurs. Victor was one of them. He not only advised his old MI5 and MI6 colleagues on technical and scientific matters, he also (according to some sources) was active in running his own agents in China and Iran. He had close connections with leading members of the Israeli government and it seems likely that he helped to set up Mossad, the new nation's Intelligence bureau . . . It, apparently, did not seem inappropriate to Britain's spymasters to share highly classified information with a man who had no official standing in the security services, which flattered Victor's ego. It enabled him to be of use to the state. A telephone call from Lord Rothschild could put an MI5 officer in direct contact with the head of a foreign legation or with the deputy head of the atomic weapons research establishment and avoid all the time-wasting of 'official channels'. A minute of Lord Rothschild and Shell scientists could be set to work on some technical problem for MI5.[3]

Lovelock was interviewed by MoD scientists at Century House, near Elephant and Castle. After the tracer was successfully trialled on Dartmoor, he was invited to work as a part-time consultant at Admiralty Materials Laboratory, near Holton Heath, which was conveniently only forty minutes' drive from Bowerchalke. There was one last courtesy by the intelligence services: double-checking with Rothschild that Lovelock could be spared by Shell.

16 June 1965, Private

Dear Lovelock

I had a talk yesterday evening with the people whom you have recently been in contact about your 'device'. They wanted to know whether I had any objection because of your association with Shell, to your advising them a bit more about your 'device'. I said no, provided it was not too time consuming. They assured me it would not be. I assured them that I would emphasise to you my hope that you would not get distracted from what might appear to be the somewhat less interesting work in which Shell is engaged. There are many problems cooking in Shell about which we shall need your help.

'Many problems' almost certainly referred to the growing evidence that petrochemical products were a threat to health, nature and the climate. Increasingly sensitive chemical detectors, including Lovelock's ECD, had opened the world's eyes to petrochemical pollution that had previously gone unnoticed, particularly the small but accumulating presence in the air of toxic vapours, such as dichlorodiphenyltrichloroethane (DDT), tetraethyl lead, freons and a host of carcinogens. These discoveries posed a growing challenge to Shell's business and Rothschild wanted to stay ahead of critics. It would not be unreasonable to assume that one of the reasons this skilled puppet master hired Lovelock was to keep the father of the

ECD under his thumb, which gave Shell a head start on environmental challenges.

So, while scientists and ecological campaigners were applying the chemical detector to try and expose industrial contamination, its inventor, Lovelock, was working with an oil baron and his secret-service associates to secretly modify the device to head off criticism and trace enemy agents. The latter was exciting work and very much in keeping with the spirit of the age. The James Bond film franchise had started three years earlier, changing the popular image of spies from shady to glamorous. Lovelock joked later that he was a 'mini-Q', in reference to the fictional secret-service boffin who equipped 007 with new gadgets before each mission. This covert work meant he had to live something of a double life. His family were not to be told. Nellie's instinct, which she confided to her diary in 1965, was correct: her son had been holding something back.

But she need not have worried about his finances. With money now coming in from Shell, MI5, JPL and the computer firm Hewlett-Packard, Lovelock was feeling flush. He upgraded his car from a Morris to a Mercedes and built a new laboratory that was equipped with the help of Shell. He was a jet-setter in a decade when this was considered a privilege rather than a burden, and he always flew first class.

'You have to think of me then as an individual earning far more money than I ever thought possible, living in a strange sort of world, where I would spend afternoons with Rothschild having been suddenly lifted from the civil service to a dreamworld,' he recalled. 'I was classed as a VIP because I was Rothschild's representative.' When he visited the Shell centres in Thornton, he was picked up by the company limousine and cowed executives would take him out on late-night benders in Welsh country pubs. 'I was treated as a mouthpiece of Rothschild, so staff wanted me to get drunk quickly. We'd go to pub after pub. These people were the top managers in the refinery.'

He was more than good value. Rothschild raised Lovelock's consultancy fee to £1,500 plus a share of revenue from inventions. Given his role and output, it was a modest sum, worth about £14,000 in today's money, and Lovelock later came to rue that most of the revenue from patents went to Shell. This was something of a lifelong complaint. Lovelock was often unwise in his financial decisions. This may have been due to his dyscalculia. But mostly, it was because his overriding priority in life was to access top-of-the-range laboratories and equipment. The acquisition of knowledge was his vice. As he put it many years later: 'There is no great sin in earning a lot of money. It's what you do with it. Luxury cars and fast women are not as good as using it to find out things.'

Rothschild realised the best way to motivate Lovelock was to indulge his curiosity and hunger for recognition. He wrote personally to the editor of *Nature*, Jack Brimble, another old boys' club connection, to ask that he publish Lovelock's paper on the detection of extraterrestrial life, which would be the precursor of later Gaia papers. He authorised Thornton to buy equipment for Lovelock's new home laboratory up to £100 (equivalent to £1,000 in today's money).[4] And, once he discovered chemical tracing would be used to detect pipeline leaks as well as enemy agents, he confidentially assigned two senior Shell scientists – Airs and Pender – to work on the new device, under Lovelock's instruction.

In doing this, Rothschild was secretly getting Shell to carry out and pay for secret-service work. Both needed a portable version of the ECD that could be carried on a helicopter, plane or other vehicle that would allow tracing over large distances. On 28 July 1965, Airs informed his boss that a miniature ECD had been made to 'Dr Lovelock's specification'. It was fitted in a briefcase. A month later, Airs wrote a memo to confirm that his team had experimented with 'discrete samples of hydrocarbons containing mixtures of organofluoro-compounds' to see which mixtures were best for the 'fingerprinting' work (for the secret service), which he distinguished from Shell's pipeline leak detection.

This had to be done in secret. If newspapers or shareholders realised Shell was aiding the secret service in this way, it would have caused a scandal. For Rothschild, who bestrode both camps, there was no conflict of interest; it was simply a matter of efficiency. He wanted to keep the project within his personal fiefdom. When Lovelock asked whether some of the research could be done elsewhere, including by 'friends' at MI5 and MI6, his boss quashed the suggestion.

14 September 1965

Dear Lovelock,
I hope you will not think the worse of me if I say that I am, I think naturally, interested in doing something for Shell and, therefore, not so keen on getting work started on both sides of the Atlantic or, indeed, anywhere apart from Shell . . . As regards going it alone with Shell, I think you know I have advised our other friends in England not to start experimental work until the end of the year and then see how far Thornton has got, to avoid unnecessary duplication and waste of money.

Within a year, the technology was sufficiently developed to enable Shell to apply for a patent for innocuously titled 'labelling materials'. Rothschild then followed his espionage instincts and covered his tracks to ensure other executives, shareholders and the media were kept in the dark about Shell's collaboration with government intelligence agencies. He told Lovelock that the veil of secrecy over the project should not even be lifted for the Shell scientist and public relations specialist Sidney Epton, who Rothschild had appointed the previous year as a point-man to work with Lovelock.

> *30 April 1966,* **Confidential**
>
> Dear Lovelock,
> There is no reason why you should not discuss with Epton the Tracer work, provided it is clearly understood by Airs and those concerned with it, and you, that in no circumstances must any reference be made to the Government work, nor to the fact that Thornton is making a gadget for the Government.

This was not the last time Lovelock was told to keep his mouth shut. The 'independent scientist' came to realise that freedom of research in the private sector was not always matched by freedom of expression. This was particularly true when the focus of his research for Shell switched from money-making inventions to costly controls on health and environmental hazards.

In the 1960s, man-made chemicals were insinuating themselves into everyday life and popular culture as never before. The contraceptive pill and psychoactive drugs made possible the flower-power revolutions of that decade. As usual, the Lovelock household was moving with the bohemian times, although more inclined towards natural narcotics. Later in the decade, Andrew, a rakishly handsome young man with long hair, was a drummer in a band and known by associates as 'one of the original Salisbury hippies'. Christine and her boyfriend would occasionally smoke marijuana in a pipe. Lovelock told his children he preferred cannabis to tobacco, and co-authored a letter to the *Times* supporting the legalisation of marijuana after the Rolling Stones singer Mick Jagger was arrested for possession.

Christine remembered this as a joyful period, if somewhat reckless. One time in Ireland, she recalled them all getting high, giggling and then deciding to buy ice creams from Uncle Don's Mountain Cabin. A stoned Lovelock then drove everyone up to the Healy Pass in great swings round the hairpin bends as they sang 'All You Need Is Love' and other Beatles tunes. The reality check followed a year later, when Andrew brought some unusually psychotropic hash to Bowerchalke that left his girlfriend traumatised and Christine suffering flashbacks for months. 'I had a terrible experience,' she recalled. 'I would get terrified on the Tube. I couldn't tell what was the real door, so I was frightened I would walk onto the track. When I picked up the phone, it turned into a banana.' Their father, typically, became curious and carried out a chemical analysis of the hash. He found it was adulterated with a powerful monoamine oxidase inhibitor, which can be extremely dangerous if mixed with other psychoactive substances or even certain foods.

Evidence of the risks posed by petrochemical products was also mounting. The problems increasingly came to dominate the meetings between Rothschild and Lovelock. They would still exchange witticisms and sometimes share a lunch. But the initially breezy relationship became strained as the topics of conversation grew more troubled. With every passing year, the petrochemical industry had more questions to answer.

Rachel Carson's landmark environmental investigation, *Silent Spring*, had been published in the United States in 1962 and its revelations about the poisoning of the countryside with pesticides were reverberating across the Atlantic. The American biologist had drawn attention to a build-up in the environment and food chain of chlorinated hydrocarbons, such as dieldrin, aldrin and DDT, which were killing birds and silencing once vibrant landscapes. The book transformed perceptions of the chemical industry and inspired campaigns against other persistent bio-accumulators that were invisible to the human eye but neurotoxic, carcinogenic or endocrine-disrupting, leading eventually to bans and other forms of regulation. *Silent Spring* is often cited as the spark that ignited the environmental movement.

Lovelock later claimed on multiple occasions that his ECD was the 'midwife' to *Silent Spring*.[5] This seems highly improbable. Science historians have examined Carson's 500 or so literature references and determined that they all used different analytical techniques.[6] However, the ECD did reinforce Carson's conclusions, and the device was increasingly used to detect other man-made toxins that were accumulating in the air, water and soil.

This led to a potential conflict of interest because Lovelock's client, Shell was the sole manufacturer of dieldrin and aldrin, which accounted for the bulk of its agricultural chemicals revenues.[7] How should an individual or a business balance conflicting environmental and economic concerns? The dilemma was spreading throughout industry at the time. Pollution problems were often first discovered by polluters, and it was their instruments that made many environmental campaigns possible. As science historians noted: 'The pesticide industry was attacked with a sword that was largely of its own making.'[8]

Lovelock claimed he met his ethical obligations by warning Rothschild: 'I alerted him to Rachel Carson's *Silent Spring*. He read the book and then on the next visit, he said: "I'm afraid she is right. We'll have to close the dieldrin plant in the Netherlands."' This was not what happened. In fact, the opposite was true. Shell maintained production and launched a publicity campaign to question Carson's scientific credentials and to reassure the public that the risks of these insecticides were overblown and the benefits of disease eradication and food production undervalued.[9] The company did not shut its plants until the US Congress and European regulators forced it to do so. Lovelock's summer home in Ireland was most likely where Shell scientists later conducted confidential measurements of aldrin and dieldrin levels in the air of Bantry Bay.[10]

Nonetheless, the public lesson Lovelock took from this – and repeated in speeches for decades afterwards – was that big corporations were not the evil, planet-trashing villains they were often painted to be by the green movement. He was convinced corporations were more far-sighted and responsive than governments because they had the resources and expertise to fully assess and respond to problems. Environmental activists, he believed, were less well informed and tended to take data out of context and exaggerate the dangers posed by minute quantities of toxins.

This ignored the reality that companies often have a greater financial motivation to hide commercial risks than a regulatory incentive to inform the public about health and environmental hazards. It was remarkable that Shell had not already revealed the impact of aldrin and dieldrin because the company had access to the best analytical equipment for pesticide detection in the late 1950s, when concerns were first raised about these chemicals.

Transparency was not a matter to which Lovelock was particularly well attuned. In public, he downplayed the possibility that corporations might cover up threats to the public. In private, he knew better. This was also true of another threat.

'One thing I discussed that could have been disastrous for Shell was tetraethyl lead. It is an extremely poisonous compound that destroys the brain. The ECD was the best way to detect it. We found the average car used to have tetraethyl lead in huge amounts from leaking carburettors. Whenever you could smell petrol, you were inhaling it. This was the only thing I did at Thornton that was not published,' Lovelock acknowledged towards the end of his life.

The history of tetraethyl lead is another illustration of the petrochemical industry's tendency to put revenue before health. In 1921, the American scientist Thomas Midgley discovered this artificial compound could prevent 'knocking', the sound produced by mis-timing combustion engines. Once lead was added to fuel, engines purred instead of spluttered. It was a chemically beautiful solution that saved petrol, made cars faster and reduced wear and tear. Unfortunately, what was good for a car engine turned out to be awful for the human brain.

Fears had already been evident at the time of the product's commercial release in the 1920s. The US Office of the Surgeon General warned tetraethyl lead was a 'serious menace to public health' that would accumulate in the environment. These concerns came horrifically true in October 1924, when five workers were killed at Standard Oil's experimental laboratories in Elizabeth, New Jersey, while dozens of others suffered serious neurological symptoms, such as palsies and hallucinations. The corporate response was brutal. A company doctor told the *New York Times* that 'these men probably went insane because they worked too hard'. Midgley blamed the workers for failing to avoid exposure of their hands and arms. Later still, industry executives insisted economic benefits outweighed environmental risks. Frank Howard, president of the Ethyl Gasoline Corporation, proclaimed tetraethyl lead as an 'apparent gift of God' that was essential for the development of civilisation. Over the next decade, thousands of tonnes were added to gasoline and released into the air when the fuel was burned.

Worries re-emerged in the 1940s and early 1950s, when the American geochemist Clair Cameron Patterson found lead in Greenland ice-cores dating back two decades, which proved the heavy metal had been accumulating across the world since its addition to petrol. In the 1960s, clinical studies proved for the first time that the compound was toxic and affected virtually every neurotransmitter system in the body. A decade later, the Pittsburgh-based paediatrician Herbert Needleman revealed higher lead levels in children's blood were correlated with decreased school performance. Lead was dumbing down an entire generation.

In the mid 1960s, Lovelock discovered a danger associated with tetraethyl lead that nobody else was looking at. Up until then and for many years after, the focus of scientists was on lead particles. But Lovelock found that there was a far greater danger posed by tetraethyl vapour. In this gaseous form, he reasoned, the toxin could pass more easily through the lungs, into the bloodstream and damage the organs, particularly the brain.

When he used the ECD to measure how much tetraethyl lead was inside the cabin of one of the old cars at Thornton, he was shocked: 'When I saw the result, I said: "Oh God, look at this!" ... From then on, I wouldn't go in cars that smelled of petrol. I told people in authority, but nothing was done about it.'

Lovelock wrote up the results in a 1967 internal paper, which revealed lead vapour levels inside a car rose steadily during a drive from Chester to London and peaked at about 50 micrograms per cubic metre, which was more than 300 times the maximum permitted industrial exposure at the time.[11]

As Lovelock pointed out in his report for Shell, those safety limits were based on lead particles, not vapour, which 'could be as much as 100 times as toxic'. Some level of exposure, he said, was possible every day for the millions who drove to work, and the risks were about to get much worse because of plans to replace tetraethyl lead with the still more volatile tetramethyl lead. He spelled out these concerns in a note in July 1967, which urged a deeper investigation. 'When gasoline is smelt some amount of lead must always be inhaled,' he observed. 'The hazard seems serious enough to justify some further thought and work.'

If Rothschild saw this, his response has been misplaced as it does not appear with other correspondence in the Science Museum archives. Lovelock never wrote up his findings for a journal: 'I never published a negative paper. I didn't see any reason to publish that, nor did Shell. It was a practical thing,' he said, adding later: 'I have a vague feeling of being slightly annoyed about being censored.'

A few years later, Lovelock mentioned his study to Derek Bryce-Smith, a chemist at Reading University, who would go on to spearhead a public campaign for lead-free petrol. It is unclear whether Jim accidentally spilled the beans in an unguarded moment, or deliberately alerted his colleague to blow the whistle, which he felt unable to do himself.

5 July 1971, letter from Bryce-Smith to Lovelock
You may remember that we had a chat a year or two ago on the subject of lead pollution, in particular the pollution

arising from leaded petrol ... I have been asked by the BBC to give a presentation of the main essence of the case by means of a lecture at the Royal Institution which will be followed by a discussion with half a dozen opponents who will be located in the audience. The whole thing will be televised ...

I recall that you referred to an investigation by a British oil company into the organic lead levels inside closed cars of various types under normal conditions of use ... I seem to remember that the tendency to release organic lead compounds into the passenger compartment was greatest with old cars, and that certain makes of carburettor were worse in this respect than others ...

I shall of course be discreet in any reference to such information. I myself am a consultant for a major company concerned with the manufacture of lead alkyls, although I may well not be so for much longer.

Lovelock's response was half-hearted. He promised to ask Shell for permission to release data, though he must have known that request would be denied.

11 July 1971, letter from Lovelock to Bryce-Smith
First I must confirm that the information on lead alkyl vapours in cars is confidential; I can and will ask if it can be released for your lecture but you can not assume that the answer will be yes.

Bryce-Smith persisted and asked again for Lovelock's permission to say he knew of 'unpublished oil company research' on the dangers of leaded petrol.

16 July 1971, letter from Bryce-Smith to Lovelock
I can well imagine that there is no chance that the company for whom you consult would agree to my making public

reference to their findings on levels of lead alkyls in cars ... I am sure that lead alkyl poisoning is much more common than anybody realises.

The archives contain no record of a reply from Lovelock. He may have been scared off because Bryce-Smith's opponent in the televised debate was a scientist from Octel, who was an associate of his colleagues at Shell Research. Shell was also a major shareholder in Octel.

In interviews for this book, Lovelock expressed remorse that he had not done more to expose the dangers of leaded petrol despite the probability that countless millions of children were being exposed to brain-destroying levels of the toxic heavy metal.

'It was bad,' he said of his reluctance to alert the public. 'I should have pushed more. I told them. I told Thornton.' His explanation for his client's behaviour showed a willingness to accept that commercial pressures trumped all. 'Shell's hands were tied. They were in a cartel with Esso and others. If a company like Shell came out against leaded petrol, it could be construed badly by other companies and civil servants.' His boss did nothing.

In his 1977 memoirs, Rothschild devotes a few pages to his favourite aphorisms.[12] The list is intended to entertain rather than enlighten, but a selection of the epigrams provides an insight into the baron's wry, even cynical, outlook:

'The optimist thinks this is the best of all possible worlds and the pessimist knows it.'   – Robert Oppenheimer.

'Two wrongs don't make a right so we had better try three.'
                                              – President Nixon

'So keen on parties that she would go to the opening of an envelope.'   – Unknown

'So many problems, so few solutions.' — R

'Let us confront this difficulty squarely, and pass on.'
— Unknown preacher

Faced with the headache of tetraethyl lead, Rothschild had allowed Lovelock to confront the difficulty squarely and then — by ignoring his findings — he passed the problem on to his successors. He attempted to replicate this tactic when evidence emerged about the climate impacts of fossil fuels, but in this case the problem was bigger, and the nagging strain was harder to shake.

It can have been no later than 1965 when Rothschild first creased his brow at reports that petrol combustion may be affecting the temperature of the planet. Fears about this had been floating around academia for more than a century. In 1824 the French scientist Jean-Baptiste Joseph Fourier had hypothesised the presence of insulating gases around planet Earth. In 1856, Eunice Foote of the United States had identified the 'warming influence' of carbon dioxide. In 1896, Svante Arrhenius of Sweden calculated how much the surface of the globe would heat up if atmospheric carbon dioxide doubled as a result of coal, oil and gas burning. Five years later, his countryman, Nils Gustaf Ekholm, was the first to compare this to a 'greenhouse' effect. Up until the middle of the twentieth century, these concerns were theoretical, which made them easy for oil, gas and coal companies to ignore as long as temperatures remained within a historic range. From the 1950s onwards, however, everyone had to reconsider their positions when the climate took an unexpected turn. The northern hemisphere started to cool dramatically — and fingers were being pointed at the oil companies. Rothschild asked Lovelock to investigate.

By this time, there was no one he trusted more at Shell.

Rothschild knew from their MI5 collaboration that he could rely on Lovelock to be discreet. More important still was that he recognised in his adviser a rare type of genius, known as synthetic intelligence – or the ability to combine different modes of thought – that he was later to laud above all others in his memoirs.

> Synthetic intelligence is called up when there is a vagueness or uncertainty or unresolved complexity about ends . . . Broadly speaking, synthetic intelligence has to invent the terms in which future analytical thinking is to be conducted, and also to invent, or to impose, new criteria of success. It is the power to make connections between disparate subject matters and problems which ordinary able analytical thinkers have not associated in their thought.[13]

Rothschild saw Lovelock in the same light, and it was why he asked him to conduct a multi-year synthesis study of the impacts of fossil-fuel combustion.

Lovelock began preliminary work on this epic commission in late summer of 1965. He arranged a series of interviews with experts in related fields and started a collection of climate clippings, which he kept on file for perpetuity:

'Northeast drought becoming a national crisis',
23 August 1965, *Chemical and Engineering News*

'Something is wrong with the weather',
17 October 1968, *New Scientist*

'Ice drifts back to Iceland', 6 March 1969, *New Scientist*

'Wintry gales greet summer's first sun seekers on French Riviera', 7 June 1969, *The Times*

'Warmer but the Thames still froze',
7 January 1970, *Evening Standard*

... And many more.

In the 1960s, fears of a new ice age were stronger than concerns about global warming, but both were manifestations of climate disruption. Based on the latest research in the United States, Lovelock suspected particulate air pollution was to blame for the fall in global temperatures, because it was masking the warming tendencies of greenhouse gases. What he had to prove was that industrial emissions were not just a local problem but a planetary phenomenon that was creating a heat-sapping atmospheric pall over large parts of the Earth.

Work on this got fully under way in 1966, when Lovelock bought a holiday cottage in Adrigole, near Bantry Bay on the west coast of Ireland. It was the ideal place to think, write and measure the chemical composition of the air to examine how petrochemical pollution was spreading across continents.

By June, the first of Lovelock's four confidential reports to Rothschild was ready.

9 June 1966, **Confidential**

'Combustion of Fossil Fuel:
Large-scale atmospheric effects.
Science fiction or reality?'
Part 1. Climatic changes

... What seems to be important is not the explanations and whether they are correct in detail but rather the almost certain fact that the climate is worsening and the probability that the combustion of fuel is responsible.

It's worth looking at that date, 1966, again. What might have been different in the years since if climate warnings had been heeded then? It would have been far cheaper and easier to deal with the causes, and early action could have prevented, or at least slowed, the melting of ice caps and glaciers, the destruction of coral reefs, the acidification of the oceans and the suffering of hundreds of millions of people from drought, flooding and extreme heat.

Lovelock's paper told Rothschild, and thus Shell, about the latest research on this problem. It explained how fossil-fuel effects had extended from local-level pollution to 'large-scale phenomena affecting the whole of the Earth's atmosphere'. As evidence for the unusual behaviour of the climate, Lovelock noted that the northern hemisphere became warmer from 1850 to 1950, most likely due to a 10 to 15 per cent increase in the carbon dioxide (greenhouse gas) content of the atmosphere. Since 1950, however, he observed temperatures had fallen back to pre-industrial levels and showed every sign of a 'precipitous further decline'.

To explain this, he cited the view of meteorologists from the US National Center for Atmospheric Research (NCAR) that both warming and cooling were caused by the combustion of fossil fuel: 'The rise, they attribute to the greenhouse effect from CO_2; the fall they believe is due to particulate combustion products, particularly ammonium sulphate, which shield the surface of the Earth from incoming sunlight, in other words an inverse greenhouse effect caused by smog.' UK meteorologists, he said, were relatively unconcerned because the temperature fluctuation remained within historic bands. But Lovelock made clear that debate over details should not diminish a clear overall message: 'The almost certain fact that the climate is worsening and the probability that the combustion of fuel is responsible.' He recommended the removal of sulphur from all Shell products.

Five months later, Lovelock submitted part two of his climate magnum opus. This second paper is breathtakingly original, mixing a deep knowledge of biology with atmospheric chemistry to make a series of bold assumptions. Here was the synthetic genius that Rothschild so admired in his protégé. It opened up an entirely new way of looking at the problem and prefigured future debates about tipping points, planetary boundaries and atmospheric science – all of which would years later become elements in the Gaia hypothesis.

16 November 1966

'Combustion of Fossil Fuel:
Large-scale atmospheric effects'
Part 2. A possible ecological effect

... The signs are that atmospheric pollution has graduated to a semi-global state ... The most disturbing thought to arise from these considerations is that with both climate and ecology, a small perturbation, itself inconsequential, can propagate to a temporarily irreversible step of great magnitude.

Lovelock started this paper with evidence that trees were growing more slowly in the cool northern hemisphere than in the south, which, he said, could be explained by the spectral distribution of sunlight reaching the surface of the Earth. The light is redder in the northern half of the planet because it is blanketed by the ammonium sulphate particles pumped out by combustion engines and industry. There were consequences of this not just for the climate but also, possibly, for the photosynthetic process of plants and other living organisms.

At this time, Lovelock and Hitchcock were already working together at JPL on the relationship between planetary life and the composition of the atmosphere. Taking this proto-Gaian idea further for Shell, Lovelock postulated that the biota (the sum of all life on Earth) had been regulating the climate, but its carefully calibrated output of gases and particulate matter was being unbalanced by emissions from fossil-fuel combustion:[14]

> The atmosphere should be considered as an integral component of the terrestrial biota with its composition normally maintained at a steady state by a system of chemical feedback loops; trace components of the atmosphere, such as methane, could participate in a regulatory capacity far more important than their slight atmospheric concentrations might otherwise suggest. The injection into such a system of large amounts of combustion products could reasonably be considered perturbing. At present the turnover of carbon by combustion in the Northern Hemisphere is 20% of the biological turnover; more important may be the turnover of sulphur which is already as great from combustion as from the biota.

He then went a step further to warn that even small changes had the potential to push the Earth system into a new state:

> The most disturbing thought to arise from these considerations is that with both climate and ecology, a small perturbation, itself inconsequential, can propagate to a temporarily irreversible step of great magnitude. From a planetary viewpoint a short-lived climatic or ecological step may be within several bounds. It should not be forgotten that such a step within these bounds is an ice age.

In his conclusion, Lovelock was cautious. As with the first paper, he did not dare to tell his client to stop pumping fossil fuels out of the ground. But he delivered another warning and urged further study.

At present the evidence linking atmospheric pollution with large-scale ecological effects is slight; nevertheless the potential for discomfort seems great enough to justify further investigation. In any event a study of pan-biotic ecology seems worthwhile in its own right. It is not only a source of disaster theories but it has the potential to predict changes beneficial to our segment of the ecology.

Rothschild was rattled. When he heard Lovelock had shared his findings with other Shell managers, he clamped down immediately. The tone of his letter – pointedly marked 'personal' – to his protégé was cold. It contained an unusual reprimand, a re-write and scepticism about the data.

25 November 1966, Personal

Dear Lovelock

I think it was a mistake on your part to give part two of your note on the possible atmospheric effects of the combustion of fossil fuels to Deacon and Izard. Neither of them are Research people and their reactions to and possible action on a document of this sort is far less predictable than in the case of a professional research man. My reaction was to get the information vetted by a top-class meteorologist, having first cleared up any ambiguities in my mind with you . . . To sum up, I do not think you ought to leave any documents with people in Shell other than those concerned with Research.

. . . Enclosed with this letter is a re-draft I did of your first document. This re-draft has not got anything omitted from it and the changes are only concerned with presentation.

Lovelock promised to check his figures at the new NCAR at Boulder, Colorado, the next time he visited the United States. But two months later, he once again invited the wrath of his boss by blithely asking if it was acceptable for him to talk to the BBC. The baron's irate reply, effectively 'mind what you say', once again highlighted the limits of the independent scientist's independence.

> 27 January 1967, **Confidential**
>
> Dear Lovelock
>
> ... I particularly don't want you to talk to non-Shell people about the subject of the note you sent me – the weather getting colder, and the cause possibly being fossil fuel combustion products in the atmosphere.

Rothschild asked for a confidential, high-level second opinion from Graham Sutton, chairman of the UK government's Natural Environment Research Council (NERC) and a former director-general of the Meteorological Office. This provided no reassurance. Sutton raised one or two technical doubts, but overall, he said Lovelock's assertion that large-scale combustion is affecting the climate was plausible: 'The suggestion is by no means to be dismissed as "science fiction", but it is impossible at the present time to give any verdict except the rather unsatisfactory one of "not proven".' He suggested Lovelock get in touch with his successor at the Met Office, John Mason.[15]

Once again his boss tapped the old-boy network to make the necessary introductions. The Met Office was then under the MoD and the senior scientists drank at Rothschild's club, the Atheneum. 'I don't think Mason would have looked at me if I hadn't come from a stable like Rothschild's,' Lovelock recalled. 'I was on a learning curve. Up to joining Rothschild, I didn't know a bloody thing about the atmosphere. I was sent to Mason. What a privilege.'

He almost blew the connection. After one particularly liquid lunch with Rothschild, Jim left a confidential letter from Mason about climate change on the train. 'I was as pissed as a newt,' he recalled. 'I thought I would be dismissed, but the letter was handed in and Rothschild simply told me I have to be more careful about boozy lunches.'

The Met Office provided him with previously unpublished surveys about the effects of volcano eruptions, which he included in an addendum to his previous papers:

24 January 1967

'Possible Large Scale Atmospheric Effects Arising From The Combustion Of Fossil Fuels'

... Volcanic dust (or any other dust of similar particle size in sufficient concentration) shall produce a powerful 'reverse greenhouse effect'.

This planted two ideas in Lovelock's mind. Firstly, it confirmed that particulate matter of all types, including dust spewed from a volcano or ammonium sulphate crystals belched by a car exhaust, can reduce the amount of sunlight that reaches the surface of the Earth. This 'turbidity' explained the fall in global temperatures and also set the groundwork for future explorations of geo-engineering techniques. Secondly and more reassuringly, it indicated the planet had temperature-shifting gears of its own – such as volcanoes – that might outweigh human impacts. This was to be a central assumption of the Gaia hypothesis in its earliest form.

Lovelock now turned his full attention to the question of 'turbidity', or airborne dust particles. His 'holiday home' in Ireland and his laboratory in Bowerchalke became bases for research into this form of pollution and the interaction between the biota and the atmosphere. He asked his daughter Christine and a local farmer, Michael O'Sullivan, to help him take measurements.

He wanted to know the origin of the late-summer haze that

blew in across the west of Ireland and southwest England from the European continent. The Met Office initially dismissed this smog as natural exhalation from the ground. But Lovelock was able to prove it was man-made pollution, most likely from holiday traffic in southern France and northern Italy. The evidence for this impressive scientific breakthrough came once again from his ECD, which measured unusually high levels of freons and other man-made compounds whenever there was a haze.

Why he was measuring these freons is an intriguing question. Lovelock has always contended that he looked for these CFCs because they do not appear in nature and are thus a clear marker of human activity. In fact, as we shall see in later chapters, this activity may well have been initiated as part of his work for the secret services.

Whatever the motive, his research provided the first proof that industrial smog was blowing across the oceans. Nuclear scientists from Harwell Atomic Energy Research Establishment were as fascinated as those from the Met Office because Lovelock's atmospheric tracing work opened up ways to study air currents, which could be useful to assess the drift of radiation from bomb tests, nuclear accidents and, potentially, wars. Lovelock published a joint paper about pollution with the two organisations in *Nature*, but public interest was low. 'Nobody paid any attention,' he recalled. 'I proved the haze was man-made. I was the only one but nobody took any notice because I was not considered a proper academic.'

He certainly had his boss's attention. 'I recall talking to Rothschild about the climate threat. It was a perennial subject,' he said. At one point, Lovelock recalled, the oil executive was so concerned about carbon dioxide emissions that he attempted to persuade other Shell directors to move away from fossil fuels. 'He said there isn't any reason why we can't be a nuclear company and said he would talk to the board. Then came back later and said: "That was a disaster. The idea went down like a dead duck."'

Unable to change the direction of the company, Rothschild revisited doubts about Lovelock's findings. Sutton and other Shell Research scientists had questioned Lovelock's figures on the amount of ammonium sulphate in the atmosphere and challenged his reliance on ice-core data to show the difference between temperature trends in the northern and southern hemispheres. Lovelock patiently rebutted these challenges one by one and followed up with a defence of his methodology which showed he was not alone in his concerns.

May 1967

'Large scale atmospheric effects'
Part 3: Stock taking

> ... The consequences if it is proven are so unpleasant that I feel strongly that at least some modest further exploration and research should be done.

In this paper, Lovelock doubled down on his findings over the previous twelve months. To strengthen his case, he compiled the key data in an updated graph and listed his sources, who included many of the world's leading meteorologists: James Lodge at the US NCAR; Graham Sutton of NERC; H.H. Lamb of the UK Met Office; A. Bratenahl at JPL in Pasadena; and Sugden, Jenkins and Windsor at Shell's Thornton Research Centre.

Lovelock acknowledged doubts – mainly raised in the UK – but insisted there was sufficient cause for concern to warrant a more thorough investigation. He then went on to suggest an examination of vegetation growth and tree rings, and a study of turbidity by the Met Office – 'if they will not do it, we should' – and historical research into visibility measurements in the British Isles going back to 1850. As an initial course of action, he repeated his suggestion that Shell should remove sulphur compounds from combustion products, which was an idea that scientists in the US were already taking seriously.

But Shell would drag out the reduction of sulphur in its petrol over decades. Meanwhile, Rothschild continued to look for holes in Lovelock's theory that fossil fuels were harming the climate. Having failed to find a strong critic in Sutton, the baron turned to one of Shell's other eminent scientists, John Cornforth, whose scratch was sharper. Cornforth lashed out at Lovelock's climate graph and Rothschild was quick to channel his scepticism.

16 June 1967

Dear Lovelock
I showed Conforth that graph, copy enclosed. He made the following comment.

'The oftener I look at this graph the more dissatisfied I become with it. What is the norm from which change is measured? . . . I think that this graph could get into the new edition of "How to Lie with Statistics" without modification, (Needless to say, this comment is JWC's.)

What is your come-back?

Lovelock was unperturbed. He replied with a barrage of bullet points. For the lay reader, his response is a tidy summary of the state of climate science in 1967.

June 1967

'Reply to JW Conforth's comment on the climate graph'

Explanation of the graph

1) The world temperature rose approximately 1C between 1850 and 1950.

2) The world temperature has fallen by about 1C between 1950 and 1965.

3) The CO_2 content of the atmosphere has increased by about 12% between 1880 and now.

4) The turbidity of the atmosphere has risen by about 30% since the turn of the century. Most of the increase is in the Northern Hemisphere.

The following are assumptions:
The rise in temperature between 1850 and 1950 was caused by the increase of CO_2 content of the atmosphere. The 'Greenhouse' effect.

The fall in temperature since 1950 has been caused by the back reflection of incident sunlight and this was due to turbidity i.e. the dustiness of the atmosphere which has increased markedly since 1950.

The CO_2 content and the turbidity are both directly or indirectly related to the combustion of fossil fuels and will continue to increase with a 7% per annum compound interest growth of production.

These two effects (CO_2 and turbidity) are the major ones controlling the future trend of temperature.

This flurry of intellectual punches marked the end of a mismatch. Cornforth was to win the Nobel Prize in 1975 for his work on enzyme-catalysed reactions, but he was out of his depth on atmospheric chemistry. Lovelock, on the other hand, was being coached by experts from the most advanced climate science institutions in the world: the UK Met Office and the US NCAR.

A couple of years earlier, Lovelock had made his first visit to this American facility to give a lecture on his Mars exploration work for NASA. The institution had opened in 1960 and, being far from the stuffy, old-establishment corridors of Yale, Harvard and Princeton, it was full of fresh thinking. Lovelock had discovered kindred spirits in the NCAR director Walter Orr Roberts and staff chemist James Lodge, who were happy to share data in return for Lovelock's pioneering research on turbidity. Their findings helped to shape the graph that Shell had been debating for more than a year and was rapidly becoming an obsession for Rothschild. The Shell executive's spymaster instincts kicked in and he encouraged Lovelock to keep close tabs on the information collected at Boulder.

6 July 1967

Dear Lovelock
How do you think it best for you to monitor the work done at the National Centre for Atmospheric Research, Boulder, Colorado to see whether that graph requires modification, etc?

Lovelock was happy to oblige, telling his boss he would make NCAR a regular stop on his future visits to the US. As a result, he said, Shell was getting the world's most advanced climate data, often before the authorities. 'I had access to NCAR and NASA climate data. Rothschild couldn't have had a better pipeline. I had access to almost everything. If you go to the lab, the scientists would tell you everything even if it was highly classified . . . Naturally I wanted all the information they had about the climate,' he recalled. Asked whether he told NCAR he was passing their data on to Shell, his answer was somewhat evasive. 'They didn't think that way. It wasn't very bureaucratic. But they would have checked that I was cleared by MI5 so they trusted me. Americans were infinitely more open.'

Lovelock visited Boulder the following November and reported back to Rothschild in a confidential memo, which strengthened his thesis that fossil fuels were causing havoc to climate systems. He ended with a typically Lovelockian left-field suggestion that sulphur should be removed from fuel and used as a currency material so that it does not re-enter the atmosphere in another form.

Rothschild's humorous reply reveals his delight at Lovelock's proof of loyalty. Thanks to his Atlantic-hopping aide, he now had eyes on the world's top climate scientists.

30 November 1967

Dear Lovelock

Many thanks for your letter and note which, as usual, greatly interested me. Powerful as the Royal Dutch/Shell Group is, I am not absolutely certain we could persuade Fort Knox to substitute sulphur for the useless metal they presently bury in it.

In late 1967, Lovelock conducted a series of experiments with a Shell scientist at Thornton, David Jenkins, that came to form the basis of his fourth and final paper for Rothschild on large-scale combustion effects. This one, written up on the final day of 1968 considers possible solutions for a destabilised climate.

The document, marked CONFIDENTIAL, was an early stab at geo-engineering. Lovelock suggested the atmosphere could be modified with a 'more efficient greenhouse gas than CO_2', such as CFCs, which could be economically distributed 'to achieve the desired climate change without overstressing the capacity of the chemical industry'. Jim lodged a UK patent for CFCs as a greenhouse gas for climate control.[16] But he knew it was a long shot. Compared to his previous three papers, this proposal was a limp effort and Lovelock knew it. The following day, he attached the remorseful letter quoted near the beginning of this chapter. It ended with a personal note of thanks to his boss for Shell's backing during the difficult early years when Jim struck out as an independent scientist and set up a family home in Bowerchalke.

Victor

1 January 1969, from Bowerchalke

Dear Lord Rothschild

... Since this is my last note to you of the present series, may I say again how deeply grateful I am for your support and encouragement during the early and critical years of re-establishment in this beloved place.

With my best wishes

By the standards of his previous correspondence with Rothschild, 'With my best wishes' was an unusually affectionate sign-off. Lovelock was remembering that Rothschild had come along just when he needed him most. The baron's support had enabled him to move out of London and buy the home in Bowerchalke that marked his escape from his mother.

This was as close as the two scientists came to intimacy in their communication. Lovelock said they did not meet socially, never discussed personal matters and he only once visited Rothschild's Mayfair home for urgent work reasons. 'We were friends but not close friends. I don't think he had anyone who cared for him much. It happens to those who reach a very high level.'

Rothschild's distance was an essential part of who he was. He was in a different social class, mixed in secret circles, and was proud of his aloof demeanour. As the baron was later to say to a gathering of British journalists: 'My cousin, Mrs James de Rothschild, who has had some experience of listening to me addressing audiences, says from time to time: "Can't you smile?" I don't think I can on occasions such as this; but I hope you will not be misled by my dead-pan features into thinking that I am not laughing from time to time at myself or even, more rarely, at you.'[17]

Rothschild's biographer suggests the baron was isolated by his status. Malcolm Muggeridge, who had known him for decades, formed the opinion that he was a man who had lost his way: 'Embedded deep down in him there was something touching and vulnerable and perceptive; at times loveable even.

But so overlaid with the bogus certainties of science and the equally bogus respect, accorded and expected, on account of his wealth and famous name that it was only rarely apparent.'[18]

Lovelock would not dream of challenging his patron or going public with his concerns about the climate impacts of petroleum products. His later public analysis of the climate issue was very different from the fears he had shared privately. In a 1971 paper for *Atmospheric Environment* he concluded breezily: 'We might find in the end that the direct aspects of combustion are the least harmful of all the major disturbances by man of the planetary ecosystem for the system may have the capacity to adapt to the input of combustion gases.'[19] Essentially, he was saying the Earth could handle whatever the petrochemical industry could throw at it – an assessment that has aged very badly.

Asked in later life whether this paper was over-complacent, Lovelock said he was 'ignorant' of the precise threat posed by fossil fuels at the time, though as we have seen he had devoted the previous six years to that precise subject, warned Rothschild of the risks on numerous occasions, and advised Shell to consider another business model. This begs the question why Lovelock had moved so far from his original alarmed position. Had he found new evidence that reassured him, or had he been swayed by his boss at Shell Research? The same question would soon be asked of the Gaia theory, which coincidently or not, emerged at exactly the same time and, initially at least, conveyed a very similar message. Lovelock said the two men talked frequently about Gaia, though there is no record of what they discussed on this topic.

Rothschild was ready to move on. Rather than an expensive solution to the climate problem, he had opted for the cheap fix of managing the fallout. He suggested modifications of Lovelock's graphs and was surely content with the soft conclusion of Jim's 'Air Pollution and Climatic Change' paper.

A couple of years later the baron gave up his post with Shell Research, quit the oil industry and moved back into government, running UK prime minister Edward Heath's 'think-tank', the Central Policy Review Staff, where he paved the way for the free-market revolution of Thatcherism by promoting a client-provider-product mode of operation at universities and in the civil service. In the 1980s, Margaret Thatcher and her ministers consulted Rothschild on matters ranging from security affairs and the poll tax to the sacking of BBC director-general Alasdair Milne and whether it was a good idea to spray Colombia's coca crops with pesticide released from planes. His discretion was admired as much as his advice. One Thatcher aide nicknamed him 'The Great Clam of Chowder'.[20]

After Rothschild's death in 1990, Lovelock described him as a 'treasured eccentric', but among his enemies he was seen as a sinister figure, a master manipulator, a harbinger of neoliberalism and a possible Soviet double agent. The latter accusation, famously levelled by Peter Wright in *Spycatcher*, was based on Rothschild's membership of the communist Apostles group at Cambridge in the 1930s and his close friendship then with Guy Burgess and Anthony Blunt, who were later outed as KGB operatives. Thatcher publicly denied there was any evidence he was a traitor, though she was somewhat wary of her adviser, who she felt had 'too much magic attached to him'.[21]

Rothschild was undoubtedly a master schemer. Many years later, Lovelock considered the possibility that his former boss had left Shell to avoid the company's growing climate jeopardy. There were other factors involved, including rows with the Dutch side of the business and political considerations, but Lovelock felt his warnings about fossil fuels surely made his boss question the future of oil. 'He was uneasy. You would be uneasy in his situation. It was threatening business.'

Almost the last archived correspondence from Rothschild to Lovelock came in 1975 after a lull of several years. The baron wanted him to check a speech he was to give on 'The Future', the very same subject he had commissioned his personal genius to think on eight years earlier.

19 August 1975

Dear Lovelock

Would you be very kind and 'correct' the enclosed address. I do not know anybody else except you with sufficient breadth of knowledge to do it . . .

If you wish to do this on a consultant's fee basis, I propose to pay in claret.

Despite the flattery and friendly tone, it was a tactical move. Rothschild needed Lovelock's endorsement as insurance. It guaranteed his former protégé would not later contradict him, claim credit or remind the world of his own vision of pollution catastrophes and a petrol-free year 2000. Lovelock was obliging.

22 August 1975

Dear Lord Rothschild

It was a pleasure to have your letter and read your prophecy . . . My problem with commenting on it is that I don't disagree with anything in it. This may be because the best prophecies are usually fiction. How can one disagree with fiction? I've always believed that fiction is more accurate though less precise than truth. It may explain why the opinions of experts are so frequently wrong . . . Finally I have checked your sums. If they are wrong, we are making the same errors.

Two months later, Rothschild delivered the speech at Imperial College. It was a masterpiece of the diversionary tactics that would later be adopted by the oil industry. There was no mention of pesticides, lead, sulphur, carbon dioxide, atmospheric pollution, global cooling, global warming, climate change or any of the other dire ecological threats he and Lovelock had identified the previous decade. There was no hint at all that petroleum companies and the burning of fossil fuels might be at all responsible for the world's problems. Instead of proposing a phase-out of the 'combustion effects' of engines, Rothschild suggested action was needed on a very different kind of explosion: the population bomb:

> 'If we assume, only of course for the purpose of illustration, that people will go on having babies at the present rate, by the year 3700 the weight of all human beings on Earth will equal the weight of the Earth. Some 1,700 years later, in the year 5400, if everyone on Earth were to be put into a hollow ball, its radius would have to be 20,000 times that of the Earth ... Some solution of the population explosion is, evidently, necessary. . .'[22]

The claim that irresponsible individuals were more to blame for environmental problems than multinational fossil-fuel companies became a staple argument for oil industry executives for decades to come. Lovelock's studies about the central responsibility of the oil industry had been passed over. But he had serenely accepted this, along with the idea that 'fiction is more accurate though less precise than truth'.

The protégé never ceased to be thankful and loyal to his patron for the support that had enabled him to leave the orbit of his mother and go it alone. Rothschild had put him in touch with the intelligence services, given him access to the most influential scientists in the UK, and opened his mind to huge thought experiments – on the climate, on the future – that made him a pioneer in what was to become the world's most important field of enquiry. More than pounds, shillings and pence, Rothschild had paid his man by liberating his curiosity. It was exactly what Lovelock craved. 'All I wanted was to do science on my own. The science I wanted to do, not what I was told to do. I just needed money to buy the apparatus,' he insisted.

Had he been used by Rothschild? 'Undoubtedly. He was probably manipulating me the whole time.'

Was Gaia also used by Rothschild to change the public debate about pollution? 'He didn't think like that.'

This was the context into which the Gaia hypothesis was born. These were the hands that sometimes rocked the brainchild's cradle in the early years. But another, altogether more essential figure was about to step in, strengthen the concept and nurture a view of Gaia that was radically different from those who would use it as an excuse to carry on as if the Earth were boundless. She would prove the best friend of Lovelock's life, arriving just in time to help him cope with a period of increasing strain. Jim would soon be torn between his old faith in industry and the new realisation of its harms. For a man who preferred not to take sides, the transition was bound to be slow and rough. But with his new friend's help, this struggle to understand prompted a semi-religious, post-humanist vision of the Earth that would inspire millions.

Lynn

*'If you will insist on probing around
in a 3 billion-year-old septic tank
you must not expect crystal clear answers.'*

On a summer morning in 1977 James Lovelock set out from his cottage in Adrigole and clambered up the boggy slopes of Hungry Hill until he reached a favourite spot close to the peak: a dry, flat rock with a sweeping view of Bantry Bay, where the richly hued landscape of western Ireland embraced the ocean and sky. There, he removed the small backpack from his shoulders, took out an apple, a chunk of cheese, a thermos flask, a notebook and a pen, and began a first draft of some of the most lyrical lines in twentieth-century science:

> The Gaia hypothesis is for those who like to talk or simply stand and stare, to wonder about the Earth and the life it bears, and to speculate about the consequences of our own presence here. It is an alternative to that pessimistic view which sees nature as a primitive force to be subdued and conquered. It is also an alternative to that equally depressing picture of our planet as a demented spaceship, forever travelling, driverless and purposeless, around an inner circle of the sun.[1]

The passage would need deciphering and revision. The scrawl was little changed from Lovelock's school years, when teachers used to complain his handwriting resembled the passage of a drunken spider after it had fallen into a bottle of ink. But the ambition of his ideas was breathtaking. He was gathering together revelations and collaborative ideas he had accumulated over four and a half decades on the frontline of science. His writing was suffused with the spirit of childhood walks with his father, the passion of trysts with Dian Hitchcock, secret conversations with Victor, a more recent collaboration with a brilliant young biologist by the name of Lynn Margulis, and his own excitement at summoning the poetic and religious muses to illuminate a theory of life on Earth. Lovelock was not just observing nature at work, he was immersing himself in it.

Working in Adrigole brought him closer to the goddess that he was invoking in an act of scientific heresy. Being in Ireland, he later wrote, 'was like living in a house run by Gaia, someone who tried hard to make all her guests comfortable':

> I began more and more to see things through her eyes and slowly dropped off, like an old coat, my loyalty to the humanist Christian belief in the good of mankind as the only thing that mattered. I began to see us all, as part of the community of living things that unconsciously keep the Earth a comfortable home, and that we humans have no special rights, only obligations to the community of Gaia.[2]

This was the moment where the theory he had conceived with Hitchcock a decade earlier took its most complete form yet, as a book: *Gaia: A New Look at Life on Earth*. This was where he began to reveal the 'cool flame' of planetary life to the world. This was the point when he first convinced himself that this exceptional planet functioned like an organism to maintain a habitable environment. There was now no doubt in his mind that he needed to go beyond Darwin – life did not just evolve to suit the environment, it had agency: 'The book was very significant for me. It established Gaia seriously in my mind. By the time I completed the book I was fairly confident I was on to a serious thing, that the Earth really operated like this.'

The publication marked Lovelock's metamorphosis from a scientist working largely in the shadows for industry into a philosopher with a public following dedicated to nature. The book propelled the Gaia theory into popular culture, inspired a new generation of scientists, influenced green thinkers and exerted an influence on earth and environmental sciences for generations to come. True to its subtitle, this was a new understanding of life that stressed interdependence, rather than competition of species. Gone were the dualities of man and nature, economy and ecology. In their places was a return – after more than two centuries' absence in the West – of a holistic, harmonious view of humanity woven into a life-maintaining planetary fabric.

But the Gaia hypothesis had not started that way. The theory's genesis was neither so benign, nor so simple. In fact, the brainchild, which had been conceived with Hitchcock during NASA's search for life on Mars, was nearly put up for adoption by Shell and other petrochemical corporations. If it had remained that way, this philosophy could have easily been remembered as little more than a public-relations ploy. That it was not is in no small part thanks to the influence on Lovelock of a new partner who entered his life at this time and would come to play a defining role in the battle for the soul of Gaia. She too became something of a mother figure.

Lynn Margulis was a biologist at Boston University, who sought out Lovelock in 1970 because – like Hitchcock – nobody else would take her ideas seriously. Although almost twenty years his junior, she had already written a groundbreaking study on endosymbiosis (a process of collaboration in which one organism lives inside the cell of another for their mutual benefit) that later won her the National Medal of Science, the highest accolade a researcher can be awarded in the United States, which helped her become one of the world's most respected biologists.

When they first met, however, Margulis was struggling to get the recognition she deserved. Those who met her never failed to recognise her extraordinary mind, and this gave her strong, important supporters but, back then, many in her field dismissed her as unorthodox or irrelevant. It did not help that she was young, untenured and a woman in a field dominated by much older men. Her book proposals and academic paper submissions were repeatedly rejected.

That was never going to deter someone as smart and as dogged as Margulis. Lynn had entered university at the precocious age of fifteen. While still in her teens, she had married the astronomer Carl Sagan, divorcing him shortly before he shared an office with Lovelock at JPL, and then raised their two children along with two others she had had with her second husband. She became an authority on micro-organisms and the evolution of eukaryotic cells. She lectured and wrote a series of profound and unorthodox papers, including an open letter in the *Psychedelic Review* that extolled the virtues of taking LSD as a means of decompartmentalising knowledge and feeling part of something bigger.[3] Outside her brief experience as a 'psychonaut', Lynn's antagonism towards artificial categories was to become a theme, as was an inclination to experience life up close and to see individuals and localities in terms of interdependent networks, or a whole that is greater than the sum of the parts.

Throughout her life, her prodigious energy was the envy of peers. 'The first thing everyone notices about Lynn Margulis is that it's impossible to keep up with her,' began a profile in *Smithsonian* magazine, which described a working day that began at 5.30 a.m. 'with a bicycle ride to her office where she does most of her work before the rest of the campus awakens':

Shared passport of Lovelock's parents, Nellie and Thomas, who travelled extensively across Europe during school holidays, leaving their son behind in the care of strangers. (Image courtesy of the Lovelock family)

'An obnoxious little boy,' as Lovelock would later describe himself as a child. His formative years were, he said, lonely but not unhappy. (Image courtesy of the Lovelock family)

A 1936 school report laments that Lovelock 'does not know how to work' in mathematics, a sign – along with a lifelong difficulty in discerning left from right – of dyscalculia. (Image courtesy of the Lovelock family)

Doodles of human faces on wormlike bodies that Lovelock called 'luvles', pronounced 'loovals'. (Image courtesy of the Lovelock family)

A page from a teenage essay, 'Goblin in the Gasworks', which prefigured a scientific life devoted to reconciling nature and industry. (Image courtesy of the Lovelock family)

Helen Hyslop married Lovelock during the Second World War and remained a loving partner, charlatan filter and pillar of support until her death more than four decades later. (Image courtesy of the Lovelock family)

A student nurse sneezes for the camera at the Common Cold Research Unit, where Lovelock conducted studies on germ transmission. (Image copyright © PA Images)

Daughters Christine and Jane had an unconventional peripatetic childhood that they reflect on differently. 'He was a genius father,' enthused Christine. Her sister would have preferred more orthodox parenting, but she appreciated his sense of mischief: 'He wasn't a boring old fart,' Jane said. (Image courtesy of the Lovelock family)

'He was a boy who wanted to catch the wind.' Jim and Helen's fourth child, John nearly died at birth and had to cope with learning disabilities during childhood. 'He was more like dad than any of us,' said sister Christine. (Image courtesy of the Lovelock family)

After the death of her husband, Nellie became emotionally dependent on Lovelock. Her diaries reveal profound loneliness and depression when her son was unable to provide the affection she craved. (Image courtesy of the Lovelock family)

Third son Andrew and girlfriend. Constant family moves and a father's anathema for formal education limited the children's academic prospects. After leaving school, Andrew joined a band, but later became a talented computer programmer and instrument maker. (Image courtesy of the Lovelock family)

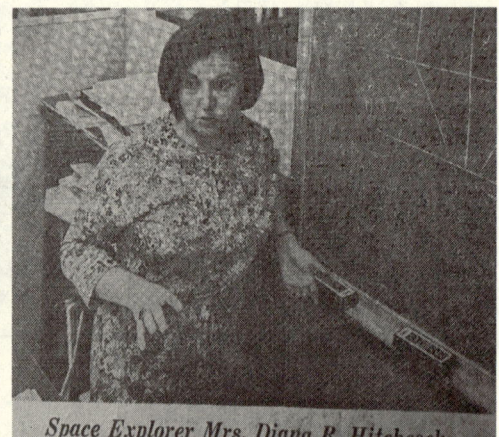

Space Explorer Mrs. Diana R. Hitchcock

The only photo in Lovelock's files of his American lover and NASA Mars program colleague Dian Hitchcock. He said she deserved 'at least half the credit' for the life detection research that was the starting point of the Gaia Hypothesis. (Image courtesy of the Lovelock family)

Victor Rothschild, the oil baron who ran Shell's research labs, mentored Lovelock in the mid 1960s, introduced him to MI6 and commissioned him to write a confidential multi-year study of fossil-fuel impacts on climate change. (Image copyright © Keystone / Hulton Archive via Getty Images)

Lovelock boozes at the Bell pub in Bowerchalke with his neighbour and future Nobel Prize for Literature laureate, William Golding. (Image courtesy of the Lovelock family)

A table of measurements from Lovelock's historic voyage on the HMS Shackleton, when he used his hand-made and self-invented gas chromatograph to reveal the build up of man-made CFCs south of the equator. This discovery made possible the discovery of the ozone hole. (Image courtesy of the Lovelock family)

US biologist and philosopher Lynn Margulis played an essential role in the formation of the Gaia Theory and counterbalancing the influence of corporate chemists on Lovelock. (Image courtesy of the Lovelock family)

Warnings of radiation risk and toxic hazard greet visitors to Lovelock's remote laboratory at Coombe Mill, Devon, where he conducted IRA bomb disposal tests and other secret experiments for the British military. (Image copyright © George Wright)

The concrete statue of a libertine that the mischievous Lovelock family named Gaia, initially as a joke. (Image courtesy of Jonathan Watts)

Romance with Lovelock's second wife Sandra Orchard began at a global spirituality conference in Oxford attended by Mother Theresa and the Dalai Lama. 'She blew his mind,' said daughter Christine. (Image courtesy of the Lovelock family)

Prime Minister Margaret Thatcher scrawls 'marvellous' in the margins of a Lovelock speech about ecological threats. She would later borrow several of Lovelock's lines for her landmark 1989 environmental speech to the United Nations. (Image courtesy of The Margaret Thatcher Foundation)

NO RAIN AND WITHOUT RAIN THERE ARE NO TREES.

GOOD CLIMATE MODELLERS FREELY ADMIT THAT THEY CAN NOT ANSWER SUCH QUESTIONS AS WHAT WILL BE THE EFFECT OF DOUBLING CARBON DIOXIDE OR OF CLEARING ALL THE TROPICAL FORESTS. THEY CAN GIVE INTELLIGENT GUESSES BUT ACKNOWLEDGE THAT THERE ARE GREAT UNCERTAINTIES. CLOUDS FOR EXAMPLE HAVE A LARGE EFFECT ON CLIMATE AND THERE IS NO WAY FOR THEM TO KNOW WHETHER CLOUD COVER WILL CHANGE IN RESPONSE TO RISING TEMPERATURES OR RISING CARBON DIOXIDE.

WHETHER OR NOT GAIA IS A TRUE DESCRIPTION OF THE EARTH MAY NOT MATTER. SOON THE EVIDENCE COMING IN FROM EXPEDITIONS AND FROM THE CONSEQUENCES OF OUR PERTURBATIONS WILL GIVE THE ANSWER. WHAT IS IMPORTANT NOW IS THAT GEOPHYSIOLOGY, THE SCIENCE THAT TAKES A TOP DOWN LOOK AT THE WHOLE EARTH, IS A SENSIBLE EARTH SCIENCE AND IS ALSO THE BASIS FOR AN EMPIRICAL PRACTISE OF PLANETARY MEDICINE.

THINKING THIS WAY HAS ALREADY ENLARGED OUR UNDERSTANDING OF THE GREENHOUSE EFFECT AND HOW IT IS REGULATED BY THE DIATOMS OF THE OCEANS AND BY LIVING ORGANISMS THAT WEATHER AWAY THE ROCKS. GEOPHYSIOLOGY ALSO LED US TO DISCOVER A GREAT NEW NATURAL CLIMATE REGULATOR, THE CONTROL OF CLOUDS BY THE EMISSION OF SULPHUR GASES FROM OCEAN ALGAE.

BUT BEYOND SCIENCE WE NEED A BELIEF IN THE EARTH AS A LIVING SYSTEM. IF YOU FIND IT HARD TO ENVISAGE OUR PLANET MOSTLY MOLTEN AND INCANDESCENT ROCK AS ALIVE, THINK OF ONE OF THOSE GIANT REDWOOD TREES THAT GROW ON THE WEST COAST OF THE USA. THEY ARE ALIVE BUT SOME ARE 3000 YEARS OLD. THEY ARE SPIRES OF LIGNIN AND CELLULOSE WEIGHING OVER 2000 TONS AND 97% OF THEM IS DEAD. THE WOOD INSIDE AND THE BARK OUTSIDE SUPPORT AND PROTECT A THIN SKIN OF LIVING TISSUE OF CELLS AROUND THE CIRCUMFERENCE OF THE TREE. JUST LIKE THE THIN SKIN OF LIVING ORGANISMS AROUND THE CIRCUMFERENCE OF THE EARTH.

IF WE RECOGNISE THE EARTH AS ALIVE WE WOULD REALISE THAT WE CAN NEVER BE ITS STEWARD. STILL LESS ITS OWNER OR TENANT. THE IDEA OF STEWARDSHIP IS BETTER THAN THE OLD TESTAMENT BELIEF IN A GOD GIVEN PLANET THERE SPECIFICALLY FOR OUR USE AND BENEFIT. BUT IT STILL IMPLIES OWNERSHIP, THE STEWARD WAS ORIGINALLY THE STYWARD THE KEEPER OF THE PIG STY. DO WE REALLY SEE GAIA AS THE PIG IN THE HUMAN STY? MY FRIEND PROF SAM BERRY HAS INTRODUCED THE TERM TRUSTEE FOR OUR ROLE WHICH IS BETTER STILL FOR IT IMPLIES A RESPONSIBILITY AND ACCOUNTABILITY TO FUTURE GENERATIONS. BUT I FEAR THAT IN MOST MINDS THESE ARE GENERATIONS OF PEOPLE.

I THINK IT BETTER THAT WE SEE OURSELVES AS MEMBERS OF A VERY DEMOCRATIC PLANETARY COMMUNITY AND REMEMBER THAT IN A DEMOCRACY, IF WE ARE IN POWER AND TREAT OUR PLANET BADLY, WE CAN BE VOTED OUT. IF WE REALLY SERIOUSLY FOUL OUR NEST WE SHALL NOT DESTROY LIFE ON EARTH, OTHER ORGANISMS WILL FLOURISH IN THEIR ENVIRONMENT, BUT WE WILL DESTROY OURSELVES.

The highest honour in geology, the Wollaston Medal was awarded to Lovelock in 2006 for his ideas, which the citation said 'gave birth to the field we now know as "Earth System Sciences".' (Image copyright © Trustees of the British Museum)

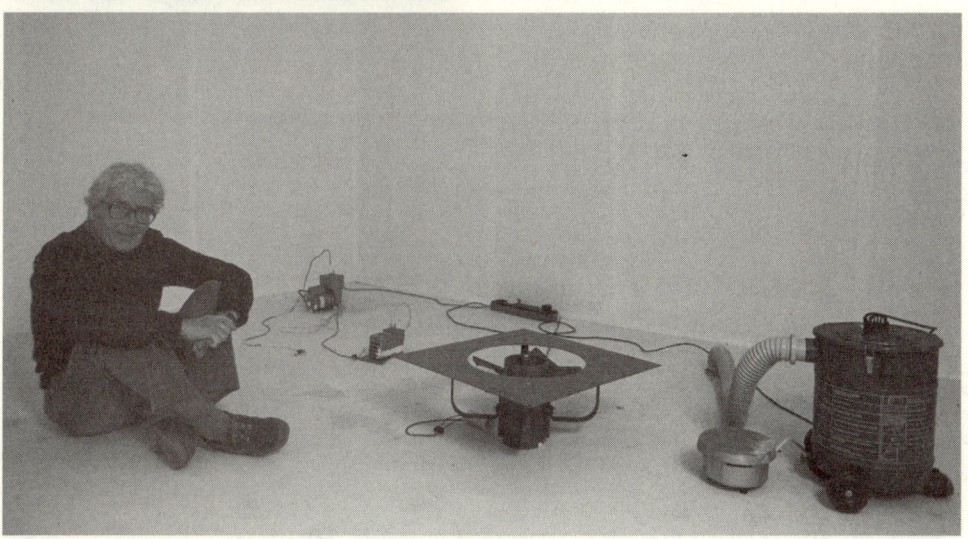

Lovelock poses for an official photo commissioned for a Science Museum exhibition of his life's work in 2015. (Image copyright © Homer Sykes)

French philosopher Bruno Latour breathed new life into the Gaia Theory just as Lovelock, then in his nineties, was starting to drift closer to climate sceptic Nigel Lawson. (Image copyright © Joel Saget / AFP via Getty Images)

Jim aged 102 at home in Matthew Cottage with the author in November 2021. (Image courtesy of Jonathan Watts)

Jim and Sandy spent their final years in this remote former coastguard's cottage on the edge of Chesil Beach in Dorset. (Image courtesy of Jonathan Watts)

The dining room where many interviews for this book took place. Among the pictures on the wall is one with a note of appreciation in green ink from 'C', the head of the British secret services. (Image courtesy of Jonathan Watts)

A sign on the walk between Lovelock's home and a nearby village. Jim was a great hiker all his life and strolled one or two miles most days even after the age of 100. (Image courtesy of Jonathan Watts)

She rarely sits still. Her speech is nonstop and, in the jargon of her profession, filled with references to DNA homologies, microtubules and antitubulin probes. She interrupts and digresses constantly because one idea triggers an avalanche of others . . . She is relentless, passionately curious, irreverent, sassy and very sharp.[4]

It was perhaps inevitable that someone deemed brilliant but rebellious would gravitate towards Lovelock. In her twenties, Margulis had focused on the unfashionable forms of life, such as bacteria, algae, worms and molluscs. While peers made careers out of providing ever more evidence for the prevailing neo-Darwinist view of life as a competitive 'survival of the fittest', Margulis spent her time trying to posit the diametrically opposite theory of cooperative symbiosis. Evolution, as she saw it, was not determined solely by a struggle 'red in tooth and claw', but by networking and interdependent relationships.

One of her theories was that atmospheric gases originated in bacteria, an idea that was dismissed by most of her colleagues. Four academic friends, including her ex-husband Sagan, advised her that the only person likely to provide a sympathetic ear was Lovelock. She reached out to him by letter for the first time in the late summer of 1970.

Lovelock, as always, was more than willing to share ideas. He informed his new penpal that he was working on his own, related theory. 'I am in the course of writing a paper on the Earth's atmosphere as a biological cybernetic system,' he wrote, reprising the ideas he had developed with Hitchcock and asserting he had also found evidence that made him 'tolerably certain that all of the thirty-six components of the Earth's atmosphere other than the rare gases and perhaps water vapour are biologically maintained'.[5] This was a stupendous claim. Mainstream thinking back then considered the atmosphere to be the product of geological processes. Lovelock reconnected it with life. He followed up by sending a draft of his paper on oxidation. Margulis was thrilled: 'I have read your oxygen article five times and finally not only do I dig it but I find it brilliant.' She, in turn, recommended that he read her work on bacteria.[6] Thus began not just a fruitful partnership but a lifelong friendship.

Margulis would make an essential contribution to the Gaia theory, providing a bottom-up explanation of how Lovelock's top-down theory might work in practice. She helped him fight battles against critics. And most important of all, she proved a critical counterweight to Lovelock's deeply entrenched fealty to industry.

Despite the formal tone of 'Dear Dr. Lovelock' and 'Dear Dr Margulis', their early correspondence revealed a shared recognition of the complementarity of their work: Lovelock staring up at the solar system to understand the biochemistry of the planetary atmosphere; Margulis peering down into the mud to grasp the metabolism of microbial life through deep time. They were equally at home with fieldwork and theory, muddy boots and pristine labs. Both saw universes: one was macro, the other micro. Together, they had the makings of a theory going back billions of years and spanning the solar system.

Lovelock invited Margulis to jointly author a paper on the atmosphere. This was the first step in what was to be an explanation of the 'circulatory system of the biosphere', which operated like a living global pump that generates, evacuates, cycles and exchanges ions and gaseous molecules, such as oxygen, carbon dioxide, ammonia and methane.

Lovelock said his partner's knowledge of bacteria filled a gap in his theory. 'Lynn was the first to tell me that we humans are huge cellular communities. We comprise 10 billion living human cells, and ten times as many more cells that are micro-organisms. There are roughly as many of these micro-organisms in your body as there are bright stars in the Milky Way.' They compensated for shortcomings in one another. Margulis admitted an ignorance of atmospheric chemistry: 'I really don't know what I need to learn or where to begin.'[7] Lovelock confided a dearth of confidence in his powers of expression. 'The evidence in favour of the atmosphere as a biological contrivance grows,' he wrote, but 'it is a devil of a job to write coherently on it. At least for me it is!'[8]

Their letters sparkle with wit and affection. Lovelock gently mocks Margulis's scrutiny of ancient detritus. 'If you will insist on probing around in a 3 billion year old septic tank you must not expect crystal clear answers.'[9] At one point in their collaboration, they chide each other for inaccuracies, prompting Lovelock to joke: I 'do agree about our sloppiness, but then you can't be an interdisciplinary disciplinarian can you?'[10]

In some respects, Margulis played a similar role to Hitchcock – a font of brilliant ideas, a sounding board, a drafter of finely tuned scientific prose, a muse, a wit and a fiercely protective mother figure. Lovelock later made the same comparison. 'Dian and Lynn were pretty tough characters,' he said. 'Both of them resembled my mother, as exceedingly strong, forthright, argumentative women. The sensible thing to do was to be fairly quiet and agree with them, but go your own way.'

Unlike with Hitchcock, however, his relationship with Margulis never strayed into the sexual. During his first visit to Boston, they worked and walked together for several days and developed enough of a rapport that Margulis, who was then married to her second husband, felt she should nip Lovelock's excitement in the bud. 'I need to make clear the ground rules. There's nothing in our relationship other than the science,' she told him in her customary straight-talking fashion. The fireworks between them were instead to be kept on a synaptic level.

In early 1972, Lovelock wrote to tell Lynn that his neighbour, the novelist William Golding, had suggested the name 'Gaia' for 'the notion of a living planet'.[11] Invoking the Greek goddess of the Earth was audacious. The idea of a living planet had been percolating in Lovelock's mind for six years, but up until that point he usually referred to it in the driest of technical language, either as a cybernetic biological system or a biologically controlled homeostatic system. For non-specialists, these names were more likely to stick in the craw than roll off the tongue. One of the great 'what ifs' of Lovelock's life is what would have happened to his hypothesis if he had stuck with these terms. Would it have found more acceptance among the scientific community? Would he have avoided the academic fights that were to follow? More importantly, would anyone have remembered his theory fifty years later?

Gaia, on the other hand, was unforgettable. In Greek mythology, she was the personification of our planet, a primordial deity whose sexual union with the sky conceived the Titans. She was the mother of all life – nurturing but also extremely powerful, sometimes resentful and capable of violent vindictiveness. In one classic tale, she plots with her son Cronus to castrate her husband with a jagged sickle while she is making love to him. In etymology, Gaia is the linguistic origin of the 'ge' in geology and geography, while her Roman equivalent, Terra, is the root of 'terrestrial'. This was a name guaranteed to stir up powerful emotions, whether awe or ridicule. It was also memorably short. Lovelock liked to call Gaia a 'good four-letter word'. Margulis sometimes described it as the 'G-word'.

Exactly when Lovelock decided on the name is a mystery. It could have been days, weeks or years before he mentioned it to Margulis. Lovelock's otherwise remarkable memory was notoriously fallible when it came to dates, which may have been another symptom of his dyscalculia. In different speeches, interviews and essays, he variously claimed he adopted the name Gaia in 'the early 1960s', '1965', 'around 1967', 'the late 1960s' and 'the early 1970s'. The latter is the most credible.

In Lovelock's archives, the first appearance of the name is on the cover of a pale-blue notebook on which Lovelock had scrawled: 'GAIA: Book 1, Ireland 1970D.' He had been friends with William Golding for most of the previous decade. The two men had struck up an acquaintance in the early 1960s, when they found themselves neighbours in the tiny village of Bowerchalke. Golding lived in Ebble Thatch, a white-walled cottage just a few minutes' walk from Lovelock's ultra-modernist home and laboratory. The novelist later wrote in his journal:

> I really must get round to writing out what I know of Jim. He is the obligatory mad scientist to be found in every village, or nearly so. He is a good deal younger than me, a cheerful illiberal liberal full of brainpower and ideas . . . When we were first here I used to drink in the pub with Jim, particularly at Sunday lunchtime and we embarked on discussions of such fantasy and bogus erudition that the pub keeper, Arthur Gulliver was very proud of us and showed us off like precious possessions of his.[12]

The landlord's pride is understandable. It was a remarkable coincidence that a remote village of barely 100 people should be home to two of the world's most influential intellects. By the mid-1960s, both had already made a name for themselves and their ideas – shared over a pint or five at the village local, the Bell – were becoming globally important. Lovelock was then working at JPL with Dian and starting to think of the life-detection ideas she had planted in his mind. Golding, a schoolteacher who studied physics at Brasenose College, Oxford, was on his way to winning a Nobel Prize for Literature. In a prolifically creative spell in the 1950s, he had published *Lord of the Flies*, *The Inheritors*, *Pincher Martin* and *Freefall*. In 1963, around the time they first met, he had just finished *The Spire* and was working on *The Pyramid*. By then, his writing was slowing down, unlike his alcohol consumption, which started to take a toll on his mental health and friendships. A decade later, the two men were still neighbours but had drifted apart, which Golding — then a depressed alcoholic suffering from writer's block — put down to his drinking: 'Probably Jim found I was too much of a boozer for him. But we still swap ideas sometimes,' he wrote in his diary.[13]

The name Gaia had cropped up when they bumped into one another on the way to the village shop. As they walked past the church and hedgerows, Lovelock sketched out his theory that the planet functioned like a living organism and Golding told him he needed a name in keeping with the immensity of the concept. He suggested Gaia. Lovelock — who had never bothered much with classics at school — initially thought he had said 'gyre'. His neighbour literally had to spell it out and explain its potency and place in antiquity. Lovelock was convinced. 'I found it attractive,' he wrote later. 'Unfortunately, the Earth and life scientists of the universities did not.'[14]

The mythical name became the subject of fierce academic debates over the next fifty years, prompting the French philosopher Bruno Latour to describe it as a 'poison gift from Golding'.[15] The term made an underwhelming debut in 1972 in a letter to the journal *Atmospheric Environment*, titled 'Gaia as seen through the atmosphere', in which Lovelock first set out his (and Hitchock's and Margulis's) theory to a wider audience: 'In this hypothesis, the air is not to be thought of as a living part of Gaia, but rather as an essential but non-living component, which can be changed or adapted as the needs require. Like the fur of a mink or the shell of a snail.'

Margulis encouraged Lovelock to take this first baby step alone. She saw it as part of the 'attitudinal change' that would be necessary for academic acceptance of a far-reaching theory. Mainstream academic journals found the idea eccentric. A joint submission about Gaia theory was rejected by the two most influential publications, *Nature* and *Science*. 'Their editorial board must be senile,' Lovelock told his partner.[16] Lynn made further revisions to what she called a 'new magnum opus' but the update was also rejected, this time by the journal *American Scientist*.

When they prepared to unveil the theory at a conference in Barcelona in 1973, Margulis felt uneasy. She was taking a risk in promoting a controversial idea when her own academic status was not yet secure. Lovelock expressed sympathy but urged his partner not to succumb to defeatism. 'Cheer up and don't be blinded by erudition,' he wrote. 'This is an area where facts are few and opinions many. Ours deserve to be heard . . . Above all Lynn do not lose heart. Gaia is no half-baked notion of a pair of amateurs to be demolished by the first glance of criticism.'[17] A few days later, he showed more sensitivity:

2 March 1973

I am concerned about the position you are in over the Barcelona meeting. I should have realised earlier that your enthusiasm was tempered with anxiety. Heaven knows I'd hate to have to prepare your stuff for a potentially hostile audience! So if you want to duck out don't fret on my behalf. It would not bother me half as much as the thought of your being savaged by the running dogs of the establishment.

In another letter, he accepted that the goddess's name might have to be scrapped if the theory was ever to see the light of day. 'By all means use any tactics to achieve its publication. I'll not fret much if you drop Gaia, its special thing for me is that it represents a potential literature citation and that it is a four-letter word.'[18]

They trimmed their ambitions and tried a small journal, the Sweden-based *Tellus*, which published a detailed elucidation of the Gaia hypothesis the following year.[19] The exposition was vivid and brave. Lovelock and Margulis argued the air, earth and living creatures are much more interlinked and interdependent than anybody had realised until then. They blurred the boundaries between life and non-life forms, and underscored the importance of understanding the workings of the Earth in a holistic way rather than breaking it down into ever smaller parts, as most scientists were doing at the time. The theory was out in the open at last. But the scientific community barely raised an eyebrow. The Gaia hypothesis 'fell like a lead balloon. It wasn't that there was any criticism, in fact there was an astonishing absence of criticism. It was just a kind of no-reaction response,' Lovelock later lamented.

The name Gaia was not the only red flag for reviewers. There were complaints about the credibility of the evidence and doubts about the reputations of the two authors: Margulis was young and had struggled to publish previous papers; Lovelock, meanwhile, had an unpaid position at Reading University that was mainly for show while much of his research for the Gaia theory was shrouded by commercial confidentiality and the Official Secrets Act, and carried out far from prying eyes on the windswept coast of County Cork.

The secluded Irish village of Adrigole, on an inner lip of Bantry Bay, had been at the centre of much of Lovelock's activities since the mid 1960s. His family first visited in 1966 on a collective whim, according to the story Lovelock tells in his memoirs.[20] But the timing and location were probably more than a happy coincidence. It was Jane, then training as a nurse, who had suggested they take a holiday. She proposed a family trip to popular tourist locations in France, Italy or Spain. Everyone was keen. But Lovelock nudged the discussion towards Ireland. He took out an atlas and, rather than looking for a beach, a mountain, a forest or some other obvious attraction, he ran his finger towards the Beara Peninsular and said enthusiastically: 'The topography looks interesting here.'

As one of the most westerly points in Europe, Beara peninsula is where sea breezes blow in from the vast expanse of the Atlantic, making it as free of atmospheric noise as it is possible to get within a day's drive from Bowerchalke. Moreover, this stretch of coastline was quiet, sparsely populated and could be reached without a passport, which meant it could be visited frequently without drawing undue attention from foreign governments. It was an ideal spot for someone who had just embarked on a programme for MI5 and MI6 to sniff the skies for chemical tracers. After the first scoping trip, Lovelock returned a few months later and bought a cottage for £3,000 cash.

For the next ten years, he and his family would drive there every summer, via the ferry across the Irish Sea. For Helen and the children, Adrigole was a refuge, where they could stroll across the hills, swim off the beach or loaf around the pool and garden (an increasingly attractive option for Helen, whose multiple sclerosis was gradually limiting her mobility). The anti-Irish Nellie never visited, which was another boon. For Lovelock, however, Adrigole was more like an immersion chamber, where he could go deeper into his work and ideas.

What started ostensibly as a family getaway soon became a working research base. After the first reconnaissance trip, Lovelock would always take his ECD – an unusual holiday accessory. Atmospheric monitoring became a bigger and bigger part of the visits. By the late 1960s, Lovelock was inviting government scientists from the Met Office and the Atomic Energy Research Establishment at Harwell to visit Adrigole and take measurements.

This is where he collected much of the world's first data on atmospheric pollution and the cycles of nitrogen and iodide. He built a laboratory next to his cottage equipped with some of the world's most sensitive atmospheric monitoring instruments. In 1975, Adrigole became the first station for a global network of atmospheric surveillance bases, funded by the chemical industry, which Lovelock initially named ALE (Atmospheric Lifetime Experiment) until more sober colleagues changed it to GAGE (Global Atmospheric Gases Experiment).

The family routine was well established. They would have breakfast, usually Quaker porridge oats with cold milk, honey and chopped walnuts, in the veranda room that looked out over the sea. Thus fuelled for the morning, Jim, Christine and other siblings or neighbours' children would go for a walk. Some days, they would wander down the drive, pass through a gate onto the land of farmer Michael O'Sullivan, who was neighbour, friend and occasional research assistant, then cross a field of fragrant gorse onto the stony path that led down to the beach. Other times, they would scramble up the rocky slopes behind their home to the 685m peak of Hungry Hill, where they could take in the panorama of Cape Clear, the Skelligs and the Atlantic. If the weather was decent, they would swim in the ocean or a mountain tarn. It was pleasant, but productive. Lovelock would take his ECD or photometer and ask Christine to jot down the readings in a notebook.

Lovelock was not just reading the atmosphere; he was writing on it. By releasing tiny trace elements of CFCs and using his ultra-sensitive ECD, he could observe how the air, thus labelled, moved across vast distances. This was not done – at least not initially – out of environmental concern; it was research for spying operations and corporate analysis. Documents in the Lovelock archive of the Science Museum reveal how he was refining his tracing equipment for at least a decade. Lovelock boasted that if an agent tipped a whiskey bottle of perfluorocarbon on the floor in Japan, he could detect it in the UK within two weeks. To do that, he needed to monitor the air in remote regions, such as Dartmoor, Bowerchalke and Adrigole.

Some topics remained murky even at the end of his life, but they raise intriguing questions. If Jim was looking for CFCs as part of a military programme, it would not be surprising that he then, perhaps by accident, started to realise that these man-made compounds were more abundant in the atmosphere than he expected, which would, in turn, have led him naturally on to his pioneering discoveries about the pollution sources of haze. When asked in later life about these possible links, Lovelock's catch-all response was that all his work was connected, and there were things he was still not at liberty to disclose due to the Official Secrets Act.

Adrigole was also where Lovelock did much of his thinking and writing, and where he started, with Margulis's help, to prove what he and Hitchcock had previously only surmised: that life on Earth — the bacteria, plankton, algae, lichen, plants, humans and other species — plays an important role in maintaining the temperature and chemical composition of the sky. Later, he would consider this his greatest achievement.

To prove this, he needed to find evidence of cycling — the circular channelling of chemicals between organisms and the air, and from sources to sinks. Following up on earlier studies by the chemist Professor Frederick Challenger, he had demonstrated in 1969 that part of Gaia's work was done through seaweed, which acted as a carrier of sulphur from the ocean to the land. He took samples of the hairy red algae *Polysiphonia fastigiata* from the Adrigole shoreline, put it in a jam jar, sealed the lid and then, when back in the cottage, extracted liquid with a hypodermic syringe and analysed it with his chromatograph. Sure enough, he found this species — and many other types of seaweed — emitted dimethyl sulphide.

With a similar procedure, he discovered the iodide cycle, which was another important piece of the Gaia puzzle, albeit less essential to life than oxygen, carbon dioxide, methane and nitrogen. He was so thrilled at his findings on sulphur and iodide that he felt someone or something had been nudging him in the right direction. 'You have two essential parts of Gaia right there. Only the ECD could've detected that. Sometimes it almost seemed as if a good fairy was looking after me,' he said later. 'I was quite excited. You could say I was comforted — "Oh yeah, nature is on my side."'

But what drove him to do this? In his memoirs, Lovelock said he was initially guided by personal curiosity: 'It is not my wont to vegetate on holiday and I began to wonder as I walked the shore what were the many varieties of seaweeds there and what interesting scientific properties did they possess.' But interviews and private correspondence suggest that even if these observations started as a private hobby, they quickly evolved into an industry-backed attempt to downplay the impact of pollution.

In an application for funding in 1971, Lovelock told Shell he would 'investigate a mechanism to account for climatic change that has so far received little attention'. Jim aimed to measure the sulphur and iodide emissions from natural sources, such as seaweed and seawater, so he could make a comparison with similar industrial gases. He claimed his preliminary findings carried out in 1970 could help the company downplay the impact of pollution: 'These observations illustrate the variety of ways by which Nature can expel unwanted materials. In many cases the rates of emissions of unwanted materials far exceed the rates at which they are emitted from man-made sources.'[21] A handwritten note on the left side of the page reads: 'Gaia hypothesis.' At this point in time, it was already evident that Gaia and industry were closely bonded.

Lovelock's exploration of the gaseous emanations of various creatures, including human industry, and his ability to trace them across the world helped him to arrive at Gaia from the bottom up, which was how Margulis saw the world. Having spent much of the previous decade considering, with Hitchcock, a top-down view of life signatures in the atmosphere, Jim now had a unique micro–macro perspective of the Earth based on close personal relationships and military-industrial sponsorship.

The British Ministry of Defence was even helping with his seaweed research. 'In those days I was working in a lab for MI5 that they had given me at Holton Heath. The Admiralty boys there were very interested in algae because they studied its effect on ship hulls. I'd ask what is this, and they would say it's dimethyl sulphide. That got me into iodine and the sulphur cycles.' He insisted this was friendly scientific advice rather than part of his wider tracing programmes for the intelligence service.

'You can see the link with MI5 but it's not in any way to do with the espionage game.'

The intelligence agencies were also involved at least tangentially with the next of Lovelock's discoveries: proof of the build-up of fluorocarbons in the atmosphere, which would lead to fears that a hole had opened up in the ozone layer – a thin belt located more than 25km above the surface of the Earth. The ozone problem would put Lovelock at the forefront of what was to become the most pressing question in the world: where were the environmental limits to industrial growth?

This debate drew Lovelock out of his depth. At the age of fifty-one, Lovelock was a one-man atmospheric monitoring operation. Thanks to his CFC tracer work for British intelligence and pollution studies for Shell and the Met Office, he had compiled some of the most comprehensive and longest-running data sets in the world. In this field, he was ahead of almost all his contemporaries, but when it came to politics, he was either too naïve, too conservative or too lacking in confidence to do anything but defend the status quo. Despite his reputation as an independent, contrarian maverick, he was embedded in the military-industrial complex. Like many scientists of his generation, he had enthusiastically embraced the petrochemical sector.

This was in keeping with his belief in the transformative power of technology, dating back to the gasworks that had enchanted him as a child, and given his father a new start in life. As he had written in his teenage essay 'Goblin in the Gasworks', he saw industry as a force for good that was entwined with nature rather than separated from it. But he was full of humility abouts the limits of science and preferred to consider himself a hands-on engineer rather than a theoretical academic. Invention was his faith, and he was proud to be an apex industrialist. He believed the planet, like the human body, was resilient to stress. Such hubris would come close to undermining the Gaia theory and it almost killed him.

Fortunately, Lovelock was being pulled in more than one direction. While his past allied him with industry, his scientific curiosity and childhood love of nature made him more aware of climate dangers and attracted him, little by little, towards a way of thinking that was less anthropocentric and more conscious of other species. Margulis encouraged him to move in this ecological direction. But his old friends from the petrochemical sector pushed him the opposite way. At the start of what would become a series of 'environment wars', they prevailed, putting Lovelock and Gaia – at least temporarily – on the wrong side of history.

Between 1971 and 1974, Lovelock pushed back the boundaries of human knowledge with a flurry of expeditions to examine whether the spread of man-made chemicals was bounded by geographic limits. He travelled higher and wider than all but a handful of atmospheric scientists before him. Curiosity, courage and a determination to follow a hunch were a big part of this activity, but it was also connected to his Gaia research and useful for his secret work for Shell and MI6. As he would often say: 'Everything I do is connected to everything else.'

This was the explorer-scientist phase of his life, when he would set out like Alexander von Humboldt or Charles Darwin to gather data in the wide oceans and high atmosphere. Unlike those gentlemen scholars of a bygone age, however, he did not have the private means to fund his travels and had to use somewhat unorthodox methods to secure passage on ships and planes.

In 1971, soon after he started to collaborate with Margulis, he had the revolutionary idea of gathering CFC data south of the equator. At the time, the southern hemisphere was almost completely free of industry, so evidence of pollution in the oceans there would prove beyond doubt that man-made chemicals from aerosols, refrigerants and other human sources were starting to accumulate across the globe.

It was a typically innovative project, but too far ahead of its time for the British government's Natural Environment Research Council, which, he said, turned down Jim's initial application for a grant. The panel of judges initially thought Lovelock was either naïve or fraudulent in claiming he could measure trace elements at a sensitivity of parts per trillion. He had to invite one of them to his lab to demonstrate he had been doing exactly that with his ECD for more than a decade.

That earned him a berth on the RRS *Ernest Shackleton*, a supply ship for the British Antarctic Survey. From the moment it set sail from Barry Docks in Wales in the late autumn of 1971, Lovelock measured levels of his beloved CFCs along the route. He had found from his chemical tracer work for the intelligence agencies that these halogenated compounds were the ideal indicator of human pollution because they were not produced by nature. He also monitored levels of dimethyl sulphide to follow up on his theories about the sulphur cycle.

The results were clear. By the time the ship reached Montevideo, Uruguay, early the following year, Lovelock had made a discovery of trailblazing significance. For the first time, he had proved man-made chemicals were accumulating not just in the soil and streams of the industrial north, but in the atmosphere and oceans on the other side of the equator. Pollution was no longer a local problem, it was global.

This should have been his greatest triumph. He had brilliantly conceived of an experiment of world-changing importance, ingeniously invented the ECD-based equipment needed to carry it out, and courageously ventured into a part of the world where no atmospheric chemist had gone before. This ought to have been the moment when he was showered with laurels. In reality, however, it was to lead to his greatest disappointment and an ethical nadir.

Close friends and colleagues recognised his achievement. William Golding predicted his old drinking partner's discovery would make a bigger mark on history than any of his own novels:

There is no curious light likely to shine down on Bowerchalke from the literateurs of the twenty first century! They are far more likely to shine down if at all, not on Ebble Thatch [the name of Golding's house] but on the house of Jim Lovelock, discoverer of the Fatal Aerosol. Maybe the sunburnt remnants of the human race in their diaspora will see him as the calamity's Ezekiel.[22]

Lovelock also shared the results with NCAR scientist James Lodge, who was so thrilled with the 'revolutionary results' that he immediately requested funds for a follow-up study from the media tycoon Robert Maxwell. Lodge said this was necessary because the British authorities failed to recognise Jim's talent: 'This could well be a matter of embarrassment for the government,' he wrote. 'He has had no difficulty in convincing me, key staff members of the Jet Propulsion Laboratory, and at the Environmental Protection Agency, and elsewhere in the US that his methods work. Only in England is he apparently a prophet without honour.'[23]

But Lovelock was his own worst enemy. He had made arguably the biggest scientific breakthrough of his life on the *Shackleton*. But, when it came to writing up the results, he failed to follow the logic of his own findings because he was too close to the industries that would be affected.

In three papers for *Nature* about the expedition, he did not fully acknowledge the environmental risks his groundbreaking study had uncovered. He proved halocarbons were building up across the globe but then insisted these man-made compounds represented 'no conceivable hazard'.[24] He later came to rue this error of judgement: 'This gratuitous blunder was due to my concern that politically minded Greens would seize on the paper as proof that the air we breathed was loaded with chlorine-containing chemicals produced by the multinational chemical industry, and that we would all be poisoned as a consequence. I should have said no conceivable toxic hazard.'

He later came to recognise he had been swayed. 'On my return from the Atlantic, scientists working for the chemical industry inadvertently drew me onto their side of the forces gathering for the Ozone War.' A more cynical interpretation would be that he had been nobbled – willingly or unwillingly – by a group that stood to lose billions from his findings and hoped to soften the blow by securing his support.

Several of the 'friends' Lovelock made in this period were chemical executives, such as Ray McCarthy of DuPont, the company that had invented freons (the trade name for fluorocarbons-11 and -12, which were widely sold as refrigerants and propellants in underarm deodorants, air fresheners, insect repellents, refrigerants, varnish sprays, oven cleaners and much else). Whether Lovelock knew what a friend should be was itself debatable. Bereft of such relationships at the start of his life, he could be slow to realise when people were using him and putting political or economic interests before truth and friendship. As Jim would admit to his family, he was susceptible to flattery.

After the *Shackleton* trip, McCarthy invited Lovelock to a chemical industry meeting in Andover, New Hampshire, with other executives and scientists to discuss the potential risks of freon. The attendees spent most of the gathering very deliberately discussing the wrong question: whether their companies' mass-produced chemicals might have a direct toxic effect. Risks of poisoning, as they already knew, were negligible so they could debate this without ever seriously having to do anything that might affect their bottom line. Meanwhile, they dodged the bigger issue.

According to Lovelock, the meeting 'only touched briefly' on the large-scale effects of CFCs in the atmosphere. He had proved they were there and in growing quantities, but Jim was predisposed to believe they were harmless. He had, after all, been using similar compounds for secret chemical tracing work for the intelligence services because he believed they were not toxic.

He had even taken out a patent on them as a possible solution to climate change. They were also essential to his atmospheric pollution studies. In a prepared speech that he jotted down in a blue W.H. Smith's notebook, he told the gathering at Andover that the best thing the chemical industry could do for this research was 'continue to make these compounds at an exponentially increasing rate' as that would be easier to analyse than a steady-state situation. The risks, he told them, were still decades off: 'Even if you can sustain such a growth rate, it would be until 2000 that they would have risen to . . . the level above which there might be some concern.' His hubris – and the company he kept at the time – blinded him to the possibility that the dangers were considerable.

Environmentalists later accused McCarthy and other members of the American Manufacturing Chemists Association (MCA) of a cover-up. The Andover gathering, they said, was an attempt to assess the economic and reputational risk of the CFC build-up and to start managing the fallout. As was the case at Shell, executives had a duty to maintain sales and profits for as long as possible, which meant downplaying threat and delaying regulation. Lovelock, who was part of this culture and thereby implicated, always insisted that his friends in the chemical industry behaved decently by looking for a practical way to solve the problem with minimal damage to jobs and the economy.

Other scientists were less naïve and more independent. A few months after Lovelock returned from the *Shackleton* voyage, word of his results reached Sherwood Rowland, a chemist from the University of California, Irvine, who was more of an academic than an industrial chemist and thus less concerned about upsetting major corporations. In collaboration with his research assistant Mario Molina, he set about disproving the common wisdom – cited by Lovelock – that CFC molecules were inert and would thus not react with any other chemicals, particularly at high altitudes.[25]

Rowland and Molina argued that propellant and refrigerant chemicals rise into the stratosphere and break down the thin layer of ozone, which shields the Earth from ultraviolet radiation. Their study, which was published in *Nature* in 1974,[26] became one of the most influential papers in the history of modern science. It won the Nobel Prize, was cited hundreds of thousands of times, provoked a political debate that would drag on for decades, and triggered a confrontation with the $8 billion aerosol industry. More broadly, this was one of the strongest signs yet that nature was unable to dilute, absorb or break down certain types of industrial waste.

This shattered Lovelock's assertion that CFCs posed no conceivable hazard, and it challenged the early iteration of Gaia as a planetary system that was so strong that industry was puny by comparison. He planned to respond as any true scientist should, by trying to disprove Molina and Rowland's lab-based assumptions with a real-life examination of CFCs in the stratosphere.

There was one snag: no commercial plane flew that high and it was not easy to find a pilot willing to take on the challenge. After the Met Office turned him down, he called in a favour with his contacts in the intelligence agencies, who arranged a seat for him on a Royal Air Force C-130 Hercules that was making a test flight from Lyneham airfield up to the plane's ceiling of 45,000 feet (13,700m). There was no cost, but his MI5 source suggested he pay the crew with a box of whisky.

Joining this test flight took courage. The Hercules was the biggest and heaviest aircraft in the Royal Air Force, and its mission was to climb higher than the plane had been before and then descend with a controlled stall. Lovelock, with his customary sense of adventure and apparent indifference to danger, was thrilled. This was just the kind of *Boy's Own* science that he had admired in his childhood hero J.B.S. Haldane.

Before take-off, he wrote to Margulis: 'I shortly have to take an aeroplane trip straight upwards to do some measurements. Let us hope that it comes down again.'[27] He took samples through a static tube in the front of the plane. The results were unequivocal. Lovelock found CFC levels fell dramatically once the plane reached the stratosphere, suggesting the compound was breaking down into chlorine atoms that attack and destroy ozone. To the credit of his scientific integrity, he published the results in *Nature*, even though they proved he had been wrong, and Rowland and Molina were right: CFCs did pose a threat.

But Lovelock continued to question the urgency and scale of the danger. He examined whether Gaia's natural sinks could make artificial CFC pollution appear insignificant. He was adamant scientists were trying to take advantage of public fears.

When Margulis told him her son Zachary was worried about getting skin cancer due to the thinning of the ozone layer, Lovelock was outraged:

15 December 1974
Your story of Zach coming home with that nonsense about fatal sunburn on account of aerosol cans is one of the most deplorable events of the year. I often wonder how someone as honest as you are can stand working with those corrupt academics who are responsible for these scare stories. Universities do seem to employ the worst of our society these days. You can tell Zach that Gaia has come to the rescue and there is nothing to worry about and that Zach can go on using his aerosol spray if he wishes. We have just discovered a huge natural source of methyl chloride . . . It means that the input of chlorine to the stratosphere from natural biological sources is probably at least 100 times and possibly 1,000 times larger than from the freons.

Lovelock made a similar argument in the US Congress the following year when he was called to testify on behalf of the MCA. Alongside him was McCarthy of DuPont, who told Congress any suggestion that freons destroy the ozone was 'purely speculative with no concrete evidence'.[28] This was untrue, as Lovelock knew from his Hercules mission. But, when he took the stand, he did not contradict his friend and client. Instead, he bought time for the chemical industry by acknowledging CFCs could disrupt the ozone layer, but insisting more studies were needed to determine at what level. Although he didn't mention Gaia, his testimony very much relied on the theory that natural sources and sinks were big enough to absorb a considerable amount of pollution.

There was no suggestion that Gaia might be vulnerable, even fragile. Lovelock was not ready to acknowledge that, even to himself. It was not how he saw the world at this point. It was, in fact, not how he saw himself. He concluded with an appeal for more time and further studies: 'My basic feeling is that we seem to be rushing precipitously into action on a rather tenuous chain of evidence.' Rowland, by contrast, argued for an immediate ban.

Lovelock's defence of the chemical industry tarnished his credibility. It left him looking insensitive and out of touch. Legislators started to crack down on CFCs and the international community later drew up the Montreal Treaty to phase out these ozone-depleting substances. The science editor for the London *Times*, Nigel Hawkes, accused Jim of being in the 'pockets of the aerosol industry'.

The barb hurt, but it was not without substance. In private correspondence, Lovelock described DuPont's McCarthy as a 'major grant aid supporter'.[29] Jim's company accounts showed that members of the MCA, including Dow Chemicals, DuPont and ICI, were a significant source of revenue between 1974 and 1983 and, in several years, they took up between a third and a half of his working time. They had paid him to study how nature produced and disposed of similar chemicals to CFCs at much larger volumes, which allowed the companies to argue the Earth system could absorb anything that industry could throw at it. In effect, they, like Shell before them, were funding Gaia studies as a potential get-out-of-jail card.

Lovelock naïvely insisted the money was unimportant: 'I'm not in the least ashamed to admit that some of my funding comes from the Manufacturing Chemists Association ... As long as they do not tell me what I should do or where I should publish I cannot see I am any more corrupted than a lawyer who defends a criminal.'[30] But as the accusations continued to circulate, Lovelock was also having to defend himself from his peers, such as future Nobel chemistry laureate Paul Crutzen who had told reporters that American scientists now privately discounted Jim's views on the perfluorocarbon problem because he was in the pocket of the aerosol manufacturing industry.

Lovelock fired off a rebuttal to the Dutchman: 'I would like firmly and categorically to state that my views and results would be no different if I were funded by the Holy Ghost.'[31] In the letter he also claimed, somewhat misleadingly, that he had worked on fluorocarbons when it was 'unfashionable in the late 1960s and with no external funding', omitting to mention he had been paid by Shell and the secret services to develop fluorocarbon tracer techniques.

Lovelock did not need to be paid. He believed what he said. In that sense, he was not 'bought'; he was naturally inclined to take the side of industry. To the extent that avarice was involved at all, it was most likely – as Rothschild had already discovered – a greed for knowledge and a desire to please. Lovelock had little interest in wealth for its own sake. He was drawn to chemical conglomerates and military institutions because they were at the forefront of science, they had an unrivalled power to investigate, and they could provide him with top-of-the-range scientific equipment and first-class flights.

These clients were effectively subsidising his Gaia studies. They helped to equip his laboratories in Bowerchalke and Adrigole, which were remarkably advanced for a one-man operation. JPL had given him the means to simulate the vacuum of space with a glass dome and diffusion pumps. Shell had authorised equipment purchases. Hewlett-Packard provided a standard gas chromatograph, which he had helped to design, along with his annual consultancy fee of £30,000. The American MCA paid for an extension to Lovelock's lab, which was built by his neighbour Michael O'Sullivan and later became the first in the GAGE global network of monitoring stations.

Believing that knowledge was more valuable than cash, Lovelock also used much of his own income on tools, including a top-of-the-range watchmaker's lathe that he used to make instruments to his own precise specifications. For him, this was a fundamental part of being a scientist like his four great inspirations – Michael Faraday, Henry Cavendish, Alan Turing and John von Neumann, who were all inventors and experimenters as well as theoreticians. 'The amount I spent on equipment was gigantic,' he recalled later in life. 'That's why I have no money now.'

From 1962 until at least the mid 1970s, Lovelock was in lockstep with oil, chemical and military interests. He felt this was where he belonged. Industry was the closest thing he had to a tribe. The executives admired him. Little wonder. For the chemical firms, Lovelock was a useful genius who could help them get ahead of public criticism, divert attention, shift the blame, muddy the waters and delay regulatory action – a playbook that was later to be used to delay action on climate change. Lovelock also offered an extra weapon. Research on the Gaia theory conveniently obfuscated the difference between man-made pollution and natural discharges of compounds. Lovelock saw this as a valid scientific question: how did the impact of man-made emissions compare with natural emissions?

Was he, then, one of the first merchants of doubt? Yes and no. Certainly he was used by CFC producers, in the same way that scientists were paid by tobacco firms, and, later, fossil-fuel companies, as an expert witness who could defend their businesses in the court of public opinion. But this does not tell the whole story. As the philosopher and science historian Sébastien Dutreuil has pointed out, Lovelock was no mere academic gun for hire.[32] He was a genuine pioneer in his field, who collected raw data and had a passion for pure science. He felt a sense of ownership over the subject and aimed at a higher goal of developing his own all-encompassing theory. He may have been used cynically, but he was not a cynic himself. If anything, friends said, he could come across as naïve.

'He is the most guileless man I know,' Stephen Schneider, a US climatologist, later told journalists.[33] It had been a very different story when they first met.[34] Schneider, who was something of an activist and idealist as well as a brilliant academic, initially considered Lovelock a scientific shill hired by industry, and his Gaia theory 'no more than an elaborate excuse to pollute'. But he was quickly convinced that Jim was genuinely interested in science, and, on the question of CFCs, wanted to avoid unnecessary job losses in the chemical industry. The two men became allies. Schneider later vouched for his English friend's sincerity: 'In the 1970s, Lovelock had two cults following him: the polluters and the eco-freaks. He did not sell out to either group.'[35]

But Jim was not always sure who he could trust and struggled to cope with the politics. 'My mind is so very occupied by this accursed halocarbon affair,' he vented to Margulis.[36] The stress was starting to get to him.

Up to this point, vulnerability was not part of Lovelock's vocabulary, either for himself or Gaia. He had spent the previous decade working at a feverish pace, commuting across oceans, smoking and drinking copiously and believing he could overcome just about any challenge. It could not last.

The first warning sign was in 1972, when he suffered the first aches of angina, curiously enough at the Andover conference.[37] He was fifty-three years old. But, as with Gaia and the threat of pollution, Lovelock was initially convinced his body was strong enough to cope and that medicine could help him push through. 'They are stuffing me full of an amazing assortment of drugs, anti coagulants – rat poison to whit – diuretics, methyldopa etc etc,' he wrote to Margulis.[38] When doctors insisted he take a rest, he used the break to catch up on his reading and to walk more, in the belief that he could exercise his heart back to health. Whenever his chest started tightening, he popped glyceryl trinitrate. 'My physician says I'm already trying to do too much. It isn't easy, when there are so many interesting things to do, to set a limit,' he wrote.[39]

Deliberately or not, he stirred Margulis's maternal instincts. Lynn had already become a font of sound advice and a soothing presence to whom Jim could vent his fears and frustrations. Sometimes she fussed over him. Other times she was affectionately scolding. Years later, friends made the obvious comparison: 'Lynn was a very strong Jewish mother figure and this triggered in Jim very mixed responses. He would sometimes liken her to his own mother, implying slightly scary and overbearing. But Lynn had a deep warmth in her that it seems Jim's mother did not. Jim sometimes found Lynn too intrusive, but Lynn was tough and could take it on the chin if Jim distanced himself.'[40]

Margulis worried over him: 'Jim, Jim, Jim . . . You love the countryside because you travel so much your life is too hectic otherwise. Please, please take care of your wonderful self.'[41] She expressed relief when Lovelock went off on another oceanic expedition, this time on the *Meteor*, because it took him out of circulation for a few weeks:

15 September 1973

Dear Jim

I'm afraid that your angina induced slow-set feedback system is out-of-whack and over compensation has led to fantastic uncontrolled activity in all directions. I'm disturbed that life is dreadful at the moment and delighted that once you get on the ship no one can get you . . . not even me. Please take care . . . If you and Helen retreat to the Welsh mountains, you can be sure that both I and Dianna [sic] Hitchcock will find you.

Love Lynn

Her warnings, like those of the doctors, fell on deaf ears. By the mid 1970s, Lovelock was working and travelling with more intensity than ever before. Alongside his usual roster of clients, he was publishing in academic journals on subjects as diverse as halocarbons, dimethyl sulphide and electron capture detection. He was heavily involved in counter-terrorism operations in Northern Ireland. He was elected Fellow of the Royal Society in 1974, alongside the grown-up baby he once held in his arms, Stephen Hawking. And the CFC-ozone debate took him back and forth between the UK, US and South Africa, on top of his chemical tracing work for the intelligence services. He barely had time to write an apology to Margulis: 'Sorry to be such a poor correspondent but when I tell you that I am averaging a visit to a different country each week I hope that you'll understand why.'[42]

Despite this exhausting workload, he and Margulis decided to double down on Gaia by taking their theory outside the unreceptive confines of academia to a wider public. Lynn collaborated with *CoEvolution Quarterly*, a magazine produced by the writer and eco-entrepreneur Stewart Brand, who had made his name with the counterculture bible, *Whole Earth Catalog* magazine. Lovelock, meanwhile, worked on an article in a popular publication in the UK. The impetus for this public relations drive did not come from either of them, but from yet another petrochemical executive, Sidney Epton of Shell, who in a later age might have been called a scientific spin doctor. Like Margulis, he was a godparent to the Gaia theory despite his very different set of priorities.

Epton had been introduced to Lovelock by Lord Rothschild in 1965. He was a Shell man through and through. The oil company had paid for his postgraduate studies at Manchester University, then taken him on at the Thornton research centre in Cheshire, where he lived and worked until his retirement four decades later. He was a fascinating character – a corporate loyalist, a trained chemist and a Jew who had joined the

Communist Party because he saw it as the best way to fight Nazism and prevent a recurrence of the Holocaust.

He was thoughtful and idealistic, with a love of literature as well as science. In a later era, he might have become an environmentalist, but at the time he entered the workforce that activity did not yet exist. Instead, he felt the best way to further himself and society was through industry. His literary flair helped him rise up the corporate ladder to a variety of roles, including de facto public relations manager of Shell's science division. He wrote up new research for the company's in-house magazine and sometimes liaised with outside journalists. Rothschild had made him Lovelock's unofficial handler and they would work together for the next ten years, becoming friends, sharing lunch in the company canteen, drinking heavily on boozy visits to refineries and visiting each other's families.

Epton was tasked with playing up Lovelock's most brilliant ideas and toning down theories that might not be conducive to the business of Shell, such as the dangers of lead in petrol. After writing his prediction of the year 2000, Lovelock started to champion the idea of a steady-state economy that would maintain the Earth's environmental equilibrium – prefiguring much later campaigns for degrowth. Epton snuffed out that line of thought in a couple of sentences that Lovelock recalled decades later: 'Jim, you must understand that without growth there would be utter chaos; the whole system of modern economics is dependent on growth.'

As a ghost-writer and editor, Epton also helped to manage Lovelock's public presentations, copy-editing speeches, writing up ideas and making suggestions. Lovelock saw him as a superior wordsmith and remembered him as the initiator of the breakthrough 1975 magazine story that catapulted the Gaia hypothesis into public prominence. 'He was a lovely man. He was the writer for lots of Shell publications. Anything interesting he'd come and talk to me and write it up. When

Gaia came along, he suggested: 'Why don't we write it up for *New Scientist*?'"

They agreed that Lovelock would sketch out the core ideas in the style of an academic paper, and then Epton would adapt this for a popular audience, and they would share the byline. Having failed to convince the academic clergy up to this point, Lovelock was more than happy to take his message directly to the congregation with the help of a superior rhetorician.

The resulting article, 'The Quest for Gaia', which was published in 1975, was given cover-story treatment by *New Scientist*. The article went considerably further than anything Lovelock had written before by suggesting the planet may be a living organism. 'This book is about a search for life and the quest for Gaia is an attempt to find the largest living creature on Earth,' the authors stated in a bold opening that echoed Hitchcock. The article continued in the same vein:

Consider the following propositions:

1. Life exists only because material conditions on Earth happen to be just right for its existence
2. Life defines the material conditions needed for its survival and makes sure that they stay there.

The first of these is the conventional wisdom. It implies that life has stood poised like a needle on its point for over 3,500 million years. If the temperature or the humidity or salinity or acidity or any one of a number of other variables had strayed outside a narrow range of values for any length of time, life would have been annihilated.

Proposition 2 is an unconventional view. It implies that living matter is not passive in the face of threats to its existence. It has found means, as it were, of driving the point of the needle into the table, of forcing conditions to stay within the permissible range. This article supports and develops this view.[43]

The authors ask whether man's activities might threaten the Earth system's viability, but by the end of the piece, readers are less concerned with the dangers than they are uplifted by the euphoric, semi-religious possibility of being part of a powerful super-organism that keeps life comfortable: 'If Gaia does exist then we may find ourselves and all other living beings to be part and partners of a vast being who in her entirety has the power to maintain our planet as a fit and comfortable habitat for life.'

Lovelock knew many scientists would be infuriated by the audacity of this idea and the purple prose with which it was expressed. He warned Margulis: 'It will probably make you wince.'[44] But the target audience was the general public. And on those terms, the article hit the bullseye. Coming three years after the first colour photograph from space showing the Earth as a 'blue marble' in a monochromatic universe, the timing was perfect. The article made the world feel more alive, even more holy, than ever before. Beyond Lovelock's wildest expectations, it took the Gaia theory to a new audience, who were eager to learn more about the planetary animal with the name of a Greek goddess.

Lovelock's letter box was flooded with media requests, book-writing offers, lecture invitations, business propositions, philosophical enquiries, environmental concerns and even (via *New Scientist*) the first communication in seven years from Dian Hitchcock, who reminded readers of her earlier role in the theory's conception.[45] The pile of mail also included a surprising number of 'crank letters of a gentle and non-aggressive kind',[46] including missives about religion – a first taste of the New Age following that Gaia was to attract.

Many of the literary flourishes and spiritual undercurrents

that made the *New Scientist* article such a hit were likely to have come from the pen of Lovelock's writing partner, Epton, who had compiled a Gaia file that included several clippings about religion.[47] Lovelock received twenty-one offers from publishers, but it was Epton who started negotiations for a deal with Oxford University Press (OUP). He informed the publisher that Shell would need to approve any book he was involved in. John Brown, the editor at OUP, was indignant: 'I would be very unhappy indeed if I were unable to be certain that we could publish the book at all until it was completed and "passed" by your employers.'[48]

Epton dropped out of the book, though Shell would continue to be a supporter of Lovelock's theory. It made business sense: the more the Earth was seen as self-regulating, the less governments would feel obliged to intervene. It is not clear if Lovelock was aware of this, but he was certainly grateful to his benefactor. When his book *Gaia: A New Look at Life on Earth* was published in 1979, he credited Epton alongside Hitchcock and Margulis as the three most important contributors and expressed gratitude to Rothschild and Shell.

By this point, Jim had taken full ownership of the theory and was nominally the sole author of the book, though there was another important, though unrecognised, contributor. An old friend, Lorna Frazer, had taken over Epton's role as a 'kind of ghost-writer'. She had been a secretary at Mill Hill in the 1950s before a career as a scriptwriter (under the pen-name Evelyn Frazer) for science fiction and horror films. Jim had been a consultant for her 1957 cryogenic drama, *The Critical Point*, for the BBC. For the Gaia book, her role was to turn Jim's 'draft chapters into a book for the general reader'. Her considerable influence on the content is evident from the fact that she was paid a third of all royalties in perpetuity.[49] Nonetheless, she received no acknowledgement (at least not in the later editions seen by this author). Jim's daughter Christine said it was also Lorna who advised her father to write the book 'like a letter to an intelligent woman'. Many of Gaia's unforgettable passages surely owe a debt to her.

The book is filled with wonder about the workings of the planet: 'Life on Earth was thus an almost utterly improbable event with almost infinite opportunities of happening. So it did.' And: 'The climate and chemical properties of the Earth now and throughout its history seem always to have been optimal for life. For this to have happened by chance is as unlikely as to survive, unscathed, a drive blindfold through rush-hour traffic.'

Other sections read like a free pass for industry: 'Could it be that the pollution is natural?' Lovelock asks. 'The very concept of pollution is anthropocentric and may be irrelevant in a Gaian context.' He also asserts that damage from factories and chemical processes is minimal: 'Ecologists know that so far there is no evidence that any of man's activities have diminished the total productivity of the biosphere.' He plays up the power of life to adapt – another trope favoured by industry: 'And aren't the living able, in return, to adapt to these episodes of pollution, just as the birch moth adapted to the consequences of 19th century industrial activities?'

The book is most compelling in its vivid exposition of how the planetary system functions, particularly the interactions between life and atmosphere. Lovelock's book was a revelation to many readers and inspired dozens of prominent academic and ecological careers, and it remained in print for decades. But it did not go down well with many established scientists. Margulis was critical, though as a true friend, she mixed candour with encouragement in her initial private assessment: 'I have nearly finished *Gaia* in print and find it delightfully amusing, extraordinarily Lovelockian, charmingly original and, thoroughly unscholarly (yes, Jim, of course for you a compliment).'[50]

Her later published take on the book pulled fewer punches and prompted Lovelock into a rare bout of soul-searching. He wrote to his partner in agony:

27 December 1981

I have now had time to read your review of Gaia . . . It has given me a bad few days, during which I had the sense not to write, but now I think that I've come to terms with it and can respond . . . It isn't easy to take from you such words and phrases as 'unreliable', 'many errors' and 'glib statements'. To me these add up to mean and dishonest like a crooked used-car dealer's advertisement. Was it really that bad, Lynn? Or is it that you felt the need then to distance yourself from that wild man who clearly had not the sense to fear the vengeance of the academic Tonton Macoute.

Margulis was too good a friend to be anything but sincere: it was part of her grounding role in their partnership. She also helped to wrench the Gaia theory out of the clutches of Shell and DuPont. While the industrialists wanted a resilient, caring Earth Mother that could rock critics into complacency, the academic insisted it was vulnerable and violent. As she put it: 'Gaia is a tough bitch.'[51] While Epton – and perhaps Fraser – were eager to tap the emotional, semi-religious power of the goddess, Margulis insisted on more academic rigour. She encouraged Lovelock to write a follow-up book that was scientifically robust, more historically contextualised and conscious of the threat posed by consumption and pollution.

For the Gaia theory, Margulis proved at least as important as Hitchcock. As well as providing the bottom-up microbial explanation of how it worked, she would later help to nudge Gaia in a more idealistic direction, championing the theory as a possible basis for a healthier and more sustainable relationship between humanity and its home planet: 'It appears we must, to survive in present numbers, adopt some version of the Gaia hypothesis: only science has the status as a belief system necessary to induce human behavioural changes on a global scale.'[52]

Free of industrial influence, she was more willing than her partner to acknowledge environmental dangers. Infinitely more socially adept and aligned politically with the progressive left, she could also see how intellectual differences over Gaia and evolution echoed and shaped wider ideological battles in society over inequality, race and gender. She associated the prevailing neo-Darwinist view of life with a 'rapacious', 'materialistic', 'destructive' 'capitalist civilization'.[53]

She and Lovelock often argued. Margulis did not accept the Earth was alive. She preferred to describe Gaia as an enormous ecosystem. Despite such differences, her evident love and admiration for Lovelock was the greatest testament to his character. She believed in him as a friend, a scientist and 'a sensitive man with a deep sense of intellectual mischief'.[54] She felt Gaia theory was 'profound and important' and arguably the major spin-off from the international space programme. In her opinion, Lovelock's greatest achievement was to meld different disciplines together to force a new way of looking at life's interaction with the atmosphere: 'For too long we have had atmospheric chemists wondering, where does all that methane come from – and biologists ignorant of where all that methane goes.'[55]

Years later when Lovelock was lying in hospital, pumped full of morphine after a urethra infection and surgery, he dropped his guard and wrote to tell Margulis how deeply he felt for her with a lucidity and directness that had previously escaped him:

> *8 October 1984*
> Lynn, I love you even though your letters make me flush like a ripe tomato ... If I am the mother of Gaia you must certainly be the father – not the midwife as you claim. It was a difficult conception I do agree – like the first child of one of those semi-fertile couples who tried and tried before nature took its course.

Together they had raised the infant Gaia. And now, with the publication of the book, they would have to protect her. That battle was about to become bruising. It was not the only war to embroil Lovelock in the late 1970s and early 1980s, and each conflict took a toll on his body and his beliefs. Despite his proud declaration to be an independent scientist and his conviction that the Earth was a self-maintaining system, it was becoming ever more clear that neither he nor Life on earth were as resilient as he had previously envisaged. There needed to be a recalibration, a period of self-reflection. It was time for hubris to meet nemesis.

Barry

'On Her Majesty's Service. Top Secret.'

Lovelock was now battling enemies on a dizzying array of fronts, none of which seemed to be going well. He was deeply involved in the Cold War and the Irish Troubles through his work for MI5, MI6 and other divisions of the British secret services. He was fighting the environmental movement on behalf of the petrochemical industry over the dangers to stratospheric ozone. And his increasingly public advocacy of Gaia theory had stirred a backlash from some of the world's most prominent academics, who duelled, in his words, with 'sharpened pens dipped in acid ink'.[1]

All of these conflicts came together and took human shape in the form of the American biologist, socialist and pioneering environmental activist Barry Commoner. Records suggest Lovelock only met him only once, and accounts of that moment differ widely, but in Lovelock's memory it left a sensation of abject humiliation. At the end of his life, he talked about that particular encounter more than any other. It heralded the most dismal decade of his life. Rightly or wrongly, he came to believe it was why he was denied a Nobel Prize.

The occasion was the 1976 Nobel Symposium, a prestigious event organised by the Swedish Academy of Sciences and influential members of the Nobel Prize Committee. The subject was human disturbance of the natural nitrogen cycle, including the impact on the ozone layer. Lovelock was among more than thirty participants, who included two old friends: Bert Bolin, who would go on to head the United Nations Intergovernmental Panel on Climate Change (IPCC), and John Postgate, a microbiologist from the University of Sussex. Also present was Dutch meteorologist and future Nobel Prize winner Paul Crutzen, and the Swedish soil ecologist Thomas Rosswall who later became Executive Director of the International Council for Science.

Lovelock believed he was being groomed for a Nobel Prize due to his groundbreaking work on the build-up of CFCs and his invention of the ECD. In late August, he was flown first class to Stockholm and put up in a plush hotel, ahead of his address in the Swedish parliament building. The presence of the science-minded King Carl XVI Gustaf only added to his sense of expectation.

Commoner was among the first to take to the rostrum to give a talk on soil and water management in which he praised the commune system of the People's Republic of China. Soon after he finished, it was Lovelock's turn to give his address on the health risks posed by photochemical smog. Compared to the febrile atmosphere in the US Congress the previous year, he felt the reaction to his talk was muted, almost too genteel. The mood of calm was misleading.

Soon after came an intervention that would rankle for the rest of his life: 'I gave my paper and sat back in my seat . . . Then others spoke and I was quiet. I guess the King must've been quite bored. Then Barry Commoner piped up: "Take no notice of Lovelock. He's a bought man of industry. He's a spokesman for them."' Lovelock could not believe his ears. The censure was nothing new. But to have this allegation levelled at him in such a setting in front of peers, politicians and a king made him want to shrivel up and die of embarrassment.

As he remembered it, the mood towards him suddenly chilled: 'Everything came crashing down. They all believed what

Commoner had told them.' Bolin, a great Lovelock supporter up to that point who had published one of the early Gaia papers, suddenly withdrew an offer for Lovelock to serve on the editorial board of his journal *Tellus*; a Swedish politician visited Lovelock's hotel room to dress him down about his proximity to industrialists; and the British scientist Postgate, who had been a regular dining companion up until that point, suddenly stopped talking to him. 'I thought he was a friend. That shows I'm no judge of character,' Jim lamented.

He blamed the repercussions of this public humbling for a series of disappointments in the decades that followed. In 1988, Postgate wrote a public takedown of Lovelock's theory, 'Gaia is too big for her boots', in the *New Scientist*. In 1995, when the Nobel Committee awarded a chemistry prize for the discoveries that led to action on ozone depletion, the laureates were announced as Sherwood Rowland, Mario Molina and Paul Crutzen, while Lovelock had to make do with a mention in the official press release of his helpful contribution to ozone chemistry. And then in 2002, Crutzen published a stinging critique of Lovelock and his theory:

> How independent was Lovelock in his early judgement of global environmental issues? I remember very well how common it was in the scientific community of the early nineteen seventies to believe that mankind could not make much of an impact on the global environment. I believe Lovelock too was a victim of this misconception further enhanced by his strong belief in the Gaia concept, as well as his close contacts with a chemical industry for which he served as an adviser (by which I do not in any way mean to suggest he was 'bought' by them). These factors resulted in an early consistent pattern of underestimating the environmental impact implications of some of his groundbreaking observations ... For his inventions and early observations of trace gases he deserves a prominent place in the history

of environmental science but it was the universities and research institutions so bitterly criticised by Lovelock that saw the real environmental implications of his measurements. Lovelock might have seen it himself if he had been more independent from industry and Gaia thinking then.[2]

Returning home after the Stockholm humiliation, Lovelock said his colleagues at Reading University wanted to sue Commoner. But he had a reason not to go public with his frustration: too many secrets might come out. Friends in the British intelligence service had told him the US activist was a hard-left sympathiser suspected of spying for the Soviet Union. For Jim, this was a comforting thought: it meant he was no longer the accused – he was a victim of a Cold War conspiracy.

Lovelock had no doubt the Soviets were aware of his work for British intelligence: 'I remember vividly one of my friends in security services took me to a window of one of their places in Curzon Street. On the other side of the road was a travel agent. He said: "That place has been taken over by the KGB and they take pictures of everyone who comes in here, so you can take it for granted that somewhere in Moscow there is a file showing your photograph."'

A Nobel Prize would have increased his status, influence and access – all of which, he reasoned, would be good for MI6 and bad for the KGB. So, he started to believe friends who suggested Commoner was part of a plot to discredit him: 'I did wonder if this statement was designed to torpedo me as an authority. The KGB were quite intellectual and they may have thought that is how they can get one over on us . . . That's what MI6 said,' he recollected. 'Barry Commoner was a card-carrying communist.'

But was he? As a prominent left-wing, anti-nuclear activist and later presidential candidate, Commoner was certainly a person of interest to the United States security authorities, but this book can reveal for the first time that the FBI was unable to prove the ecological campaigner was a Soviet Spy despite monitoring him for more than thirty years. Freedom of Information requests to the FBI and Commoner's archive in the Library of Congress suggest the biologist was simply a scientist with a strong social conscience.

Commoner was a contemporary of Lovelock's and, like him, a scientist who believed knowledge could improve the world. Both had a holistic vision of the environment and felt it was impossible to understand how the world worked by studying only one discipline. That is where the similarities end. Their childhoods, ideology and views on industry, the military and transparency were in opposition. While Lovelock was a lonely child who grew up in fusty, conservative Britain, Commoner was raised in the lively social hothouse of Brooklyn by Russian-Jewish refugee parents. Social justice mattered more to him than scientific knowledge. Every discovery, he felt, should have a social purpose.

During the Second World War, he had served in the US Navy on bio-military projects. He started to question this work after developing a pesticide-spraying device that unintentionally led to the destruction of marine life and the spread of disease among soldiers. The lesson he took from this was that humans cannot interfere with one part of an ecosystem without triggering a reaction elsewhere.

He also started to appreciate the dangers of weakening scientific ethics and tightening scientific secrecy, which would be themes of his later activism. As he cautioned in an unpublished paper titled 'The Scientist and Political Power', the integrity of science was 'the sole instrument that we have for understanding the proper means of controlling the enormously destructive forces that science has placed at the hands of man'. Should the integrity of science be eroded, Commoner warned, 'science will become not merely a poor instrument for social progress but an active danger to man'.[3]

As a young academic after the war, his groundbreaking studies on plant physiology won high praise from peers and earned him a professorship at Washington University in St Louis, though his socialist politics, confrontational style and criticism of the military-industrial complex gave him a reputation as a troublemaker in more conservative circles. This intensified in the 1950s, when Commoner led opposition to nuclear weapons testing by showing that fallout could be detected in baby's teeth around the world.

In that era of McCarthyite red scares, he inevitably drew the attention of the FBI, then led by the rabid anti-communist J. Edgar Hoover. The first record in Commoner's FBI casebook on 5 February 1951 noted his contact details were found in the address book of 'known Soviet espionage agent' Israel Halperin, who was accused of violating the Canadian Official Secrets Act. Thereafter FBI agents kept up intermittent surveillance, talking to his work colleagues, students and neighbours but never finding anything that proved he was anything other than a civil rights idealist with a talent for riling opponents. An FBI report on 19 October 1956 cites a source who 'received a strong impression' that Commoner may be associated or sympathetic with the Communist Party, but then admits he has nothing on which to base his assumption. This continues intermittently for decades, with insinuation but no proof that he was anything but a feisty activist.

Commoner often used academic gatherings to publicly confront those he disagreed with. At the first international meeting on the environment – the 1972 Stockholm Conference – he interrupted an address by the demographer Paul Ehrlich who he labelled 'anti-human' for expressing the view that population growth was responsible for the world's ecological problems.

After nearly fifty years, the spicy encounter that Lovelock remembered at the 1976 conference is impossible to verify. There was, not surprisingly, no mention of any hostility in Bolin's official write-up of the event for the Royal Swedish Academy of Sciences. Bolin died in 2007, Commoner in 2012,

Postgate in 2014 and Crutzen in 2021. Indirect enquiries made to the Swedish king suggest he has no memory of Lovelock's humiliation. Thomas Rosswall, one of the other surviving witnesses, said he cannot recollect any confrontation. Anders Wijkman, who was a Swedish parliamentarian at the time, also has no memory of the incident and disagrees that the Nobel Symposium was seen as a grooming event for a prize. What is undoubtedly true, however, is that Lovelock felt wounded.

The public denunciation – or, at least, Jim's impression of it – struck to the core of his identity. Ever since 'Goblin in the Gasworks', he had tried to reconcile industry and nature - that was a noble aim of his science. Yet, the purity of those intentions was being challenged. He was being labelled 'a bought man of industry'.

Commoner was not the first to make that accusation, but his words had extra force coming from an icon of the environmental movement. This was an emblematic challenge in an age when industrialists everywhere were being forced to answer difficult questions. Like many of them, Lovelock's first instinct was denial, then suspicion of his accuser's motives, then broad resentment towards green activism; but Commoner – and other accusers – had also stirred a germ of self-doubt.

In public, Lovelock remained loyal and grateful to industry, but in private, he started to put more distance between himself and his friends in petrochemical companies. He never stopped blaming Commoner for frustrations, disappointments and stress. But he eventually acknowledged a truth he had earlier refused to admit: 'We're all bought men in an industrial society. It was unfair to single me out.'

Real or imagined, the furore also touched on a deeper question of belonging that may well have gone back to his childhood. For all of Lovelock's self-identification as an 'independent scientist' and natural loner, he craved recognition from intellectual peers. This led to a tension between his instinct to rebel from one establishment – the academic elite that had rejected Gaia papers and overlooked him for a Nobel – and his willingness to embrace the still more powerful and secretive military establishment. He relished being part of that club and was fiercely loyal to his comrades in arms.

Back then, he did not think about it in those terms. There was little time for self-reflection in the late 1970s and early 1980s. Lovelock was at war. Britain's MoD increasingly needed him as a consultant in Northern Ireland. This was hands-on work, often carried out at some personal risk in the countryside of Armagh and County Tyrone – a far cry from the more cerebral and remote research Lovelock had been doing for the intelligence services in the 1960s. Once again, he depended on the 'chemical sniffer' abilities of the ECD, which could be applied to detection of explosive caches belonging to the Provisional IRA or materials that could be used for bombs. 'We went around the countryside,' Jim recalled. 'They used ammonium nitrate fertiliser and anything combustible like diesel. That was a favourite.'

The first bomb of the Irish Troubles had gone off in 1969. In the following three decades, there would be another 16,208 more explosions, which claimed the majority of the 3,387 lives lost in the conflict. At first, the British Army had no expertise in dealing with such attacks. Ammunition Technical Officers (ATOs), whose job had previously been to inspect munitions, were initially dispatched to dispose of the bombs with little more than a Stanley knife and a pair of tweezers. Many died and their success rate was low.

During those early days they managed to defuse only one in ten bombs that were planted. In the mid 1970s, there would be a thousand bombs every year and more than a dozen ATOs were killed in the line of duty. Military chiefs on the ground realised they needed outside expertise. One former ATO, Dave Greenaway, recalled: 'We had scientists in the UK working around the clock to produce things that could assist us and they came up with disrupters. Disrupters were normally fired by explosive and sent something into the device so fast that it would cut wires quicker than the battery could send power to the detonator.'[4]

With both sides constantly updating their techniques, Northern Ireland became a laboratory for bombmaking and bomb disposal technology. Lovelock was part of that cat-and-mouse contest. 'I worked out methods to defuse detonators. It meant that I had to handle explosives,' he recalled. This was an area where he could give full rein to his inventiveness. Over the years, the British Army used a variety of outlandish devices – boot bangers (water jets that break circuits), steam and acid liquefaction, wheelbarrows (a remotely controlled though very rudimentary robot) and foaming pigs (foam squirters that diffuse blasts). Whenever they developed a new disruption kit, Lovelock or one of his colleagues would have to go to Belfast to explain how it worked.

Many of the strategies were later used to counter improvised explosive device (IED) threats in Afghanistan, Iraq and other hotspots. When asked if he had saved many lives, Lovelock replied: 'I hope so. I can't go into the techniques because they might be used again. But it was daily work. They're tireless, the terrorists.'

As an obsessive bombmaker since childhood, Lovelock told his superiors that if they really wanted to end the conflict, he could rig up a booby trap to one of the IRA caches which would be so powerful that it would take out large numbers of the enemy in one go. This remarkably un-Quakerlike suggestion was dismissed by his superiors as excessive.

Friends said he loved this work more than any of his other activities. Lovelock was never a full-time member of British Intelligence, but he was one of its most senior consultants. He said that up until the age of 100 his company, Brazzos, received cheques from Coutts bank for this work. And he was in frequent contact with secret-service officials, who he referred to as the 'purple people'. 'I have a pretty good idea of what it's like to be a spy. I suppose I was a spy in some ways,' he said. 'My rank? I was a visiting professor. On the ship *Vengeance*, I was in charge of a group of commanders. In the lab at Holton Heath, there was always a civil servant or a military officer nominally in charge, but they would defer to me. I thought the system regarded me as someone they could trust. That didn't bother me as long as the money flowed. All I needed was enough money to buy equipment and keep the family in food.'

At the Admiralty scientific research unit in Holton Heath, a source confirmed: 'Jim didn't have a rank but he was running the show. The work was mainly trying to stop the bad guys from blowing things up. But there was also tracing work to trace people.' Lovelock's team labelled targets with fluorocarbons that could be tracked – up to a point – by the ECD. 'The Russians knew that they were tagging them with radio waves but they knew nothing of the fluorocarbon markers on their backs,' Lovelock said. The technology was novel but inaccurate. 'If the wind was blowing in the wrong direction, you couldn't pick it up, so it was not entirely successful, but we had a lot of fun.'

On one earlier occasion, Lovelock was asked to help MI5 track a person who was suspected of stealing sensitive information. The target had been chemically labelled and, to monitor his movements, an adapted ECD had been placed inside a box marked 'University of Reading pollution study' on a road where the target was thought likely to pass in a Russian-owned car. But it was a clumsy operation. The box was in an all too conspicuous location outside of a school in East London. Local residents became suspicious. 'They started asking us: "Why are you spying on us?"' Lovelock recalled. Someone reported the incident to the University of Reading, where Lovelock was registered as a visiting professor. He had to apologise and explain what had happened to his department boss, Peter Fellgett, after which the incident was hushed up.[5]

Lovelock's family had long known he was working for the MoD. Helen, Christine and Andrew were directors of Brazzos and were gradually let into more details of their father's activities. They held annual meetings to discuss the accounts, which were always bolstered by MoD contracts. Christine jokingly asked her father whether the letters between them should be read 'under a fluorescent light just in case'. But they were never sure of the full extent of his work.

So it came as a shock to them when Lovelock decided to move from their beloved Bowerchalke. The ostensible reason – given in Lovelock's memoirs – was that he had grown frustrated with newcomers from London who wanted to turn the community into a twee model village. He had also come into a substantial sum of money through the sale of a patent for a version of the ECD to Dow Chemicals, a year after he had testified on their behalf to Congress. This enabled him to afford an upgrade. He purchased a farm in Dorset for £38,000 in March 1977. The nearest neighbour to this isolated plot was miles away. Jim had finally found what he had dreamed of in 'Goblin in the Gasworks': a secluded refuge where he could devote himself to science.

After an early visit, Christine described it in her diary as follows:

1 March 1977

This new place we went to on Saturday is called Coombe Mill. It is about 4 miles across the hills from Launceston and is actually just inside of Devon although Launceston is in Cornwall. It is approached up a little lane that is private after half way – with a gate across the road – and has 13 acres, divided into about five small very green fields with hedges. The house is not very big but it is built on a bluff of rock overlooking a salmon river and a sunny southerly aspect.'

She, Jane and Andrew feared the impact of the move on their mother, who had to leave her support network in Bowerchalke even though her multiple sclerosis was advancing. 'I didn't like Coombe Mill much,' Jane said. 'My gut feeling is that it was a dreadful place for a woman to be isolated in. It was the wrong place for Mum and maybe also for Dad. It was a long way from there to London.' As always, Helen made the best of things. On the day of the move, she celebrated by baking a cake and Lovelock popped a bottle of champagne.

Given the timing, the move to such a remote area was surely connected to the ever more secretive and dangerous bomb disposal work he was being asked to do in his home laboratory. He had become the freelance risk-taker of first resort. 'The reason I worked with detonators and explosives at quite a scale at Coombe Mill was they [in the MoD] weren't allowed to do it any more. There would be a discussion and someone would say: "Call up Jim, he'll do it."'

A secret-service colleague said Lovelock treated the new home as a scientific monastery: 'He was happy to use Coombe Mill as a test ground for things if needed. He did experiments there that wouldn't be allowed on an official site . . . He disliked bureaucracy and just wanted to get on with things. He always used to say that it was easier to get forgiveness than permission.' The colleague described Lovelock as a substantial figure in the protection of the nation, though few people were ever aware of it. 'What he did in reference to terrorism was very important. It contributed to saving lives. No question.'

Lovelock was thrilled by the intrigue: 'A motorbike would pull up with a soldier in full uniform delivering a brown envelope marked "On Her Majesty's Service. Top Secret". I was living in a film drama all the time. It would be a red wax seal on the back of the envelope.' On one occasion a military helicopter landed outside the house to collect him for an urgent mission.

He converted the old barn into a laboratory, where he would also store volatile materials. Over the years, he worked with nitroglycerin, Semtex and radioactive materials. To deter stray hikers and nosy neighbours, there was a radiation warning sign at the gate beside the name 'Coombe Mill Experimental Station'. A little further on was another bold sign, 'Exponential Dilution Chamber', which Jim joked would discourage snoopers since it sounded vaguely sinister, as if an intruder might get exponentially diluted. Trucks would deliver large quantities of explosives. 'If the whole lot had gone up, it would've blown up the house ... They told me if you have any problems don't call us, we never delivered this. If the police had come, I would've been jailed.'

Several years later, Lovelock showed his lab to a young scientist named Tim Lenton, who would go on to become his PhD student and collaborator. Lenton recalled his astonishment at the contents of the chemical trove: 'He opened an old ice-cream tub to reveal an orange, puttylike substance, asking me what I thought it was. "Plasticine?" I ventured. "Semtex!" he replied.'

Along with risk, the work necessitated secrets and lies. Lovelock had to keep certain activities from his family. 'I wouldn't have dreamed of telling my mother that I worked for MI5 and MI6. And I didn't tell Helen and the kids. I was living quintuple lives,' he recalled. 'Never tell anyone anything. It's all part of the game.' In public, he tried to avoid the subject of Northern Ireland and would occasionally dissemble to throw people off the scent. In an essay written for the *Journal of Chromatography* in 1981, he boasted that his ECD could be used to find explosives, and then added untruthfully: 'I was glad not to be personally involved in this new application because of close friendships and associations in Ireland.'[6]

He later acknowledged there was a dirty side to the work that he did not dwell on at the time. 'It was terrific fun and terribly exciting,' he said. 'I think I am just a kid in some ways still, who has never grown up. Also maybe I felt a guilty conscience about being a conscientious objector ... I always thought that if you think something is right and you are not doing it for obviously selfish reasons, go and do it. It is a male thing. I was putting the family at risk doing all sorts of silly things in a way. It never occurred to me.'

The family continued to visit Ireland for several months every summer. His friends warned him that Adrigole was a holiday resort for IRA hard men: 'I was living in a house in Ireland for a third of every year doing climate research. I was bumping into IRA activists all the time,' he recalled. 'Finbar O'Toole, the postmaster, was the alleged leader of the local IRA. I had to be careful about letters. And I was worried about phone calls because his daughter ran the telephone exchange . . . My greatest fear was that someone [in the British secret service] would call me and the IRA postmaster at the exchange would hear.'

In the summer of 1979, Lovelock was shaken by news that the Provisional IRA had assassinated Louis Mountbatten, the great grandson of Queen Victoria and a former head of the British Armed Forces, while he had been on holiday in County Sligo, Ireland. That bomb also killed Mountbatten's young grandson, daughter-in-law and a local boy. For Lovelock, it was all too close to home.

Five days later Jim confided his worries in a letter to Margulis:

1 September 1979
Next week Lynn we leave for Ireland for two weeks. I must reluctantly admit for the first time that I have doubts about the wisdom of going there. I am sure that our many friends in County Cork will welcome us as ever and no hostility to us as representatives of our tribe will be shown. But we have been warned that as a consequence of my work on bomb detectors that we are potential 'legitimate targets' for the IRA.

Fear had never previously been part of Lovelock's life – he had always behaved as if he were indestructible. But now that he had entered his sixth decade, sickness, injury and death appeared to be stalking him. His mother was suffering dementia. Lovelock could not cope with seeing such a sharp mind so ruinously blunted so he stopped visiting, and his letters to her dried up. The sad woman's last diary entries were filled with self-reproach after a row with her son: 'Never before has Jim ever spoken to me as he did today. He has finished with me and will never come and see me again on his own.' After this, Nellie moved to a care home for the blind in Plympton, a suburb of Plymouth. Christine took her new daughter Jane for a visit, noting poignantly: 'Nell didn't know how to hold a baby.' Her long, lonely life ended in August 1980.

The family believed Lovelock would soon follow. On New Year's Day 1981, he overturned a tractor at Coombe Mill, became jammed against the steering wheel and damaged his left kidney so badly that it never functioned again. Then his heart problems took a turn for the worse. His personal and highly unorthodox treatment for high blood pressure and a blocked coronary artery was to frequently walk the 2-mile round path from Coombe Mill to Emsworth, which included several steep hikes. He would pop trinitrin every half a mile to keep the angina at bay, but he was pushing too hard. On several occasions he collapsed and once woke up a few minutes later to find himself lying in the middle of an empty country road near Dartmoor: 'The pain was getting too bad. I lost consciousness. I did have a lot on my mind.' For the first time, he felt fragile. 'My body told me and, I half understood, that death was close, and it put me in a strange frame of mind.'

Helen continued to work full time as his personal assistant, but she too was in need of greater care. She was not bedridden but needed a mobility scooter to tend to the garden and, later, a stairlift to help her up to the first-floor bedroom. More arduous domestic tasks and nursing duties were left to her son John and a kindly neighbour called Margaret Sargent. Helen and her husband no longer had a sexual life together and Lovelock looked for partners outside his marriage. One woman in particular came to play an important part in this painful phase of his life, helped him to face his own fragility, and may even have made an indirect contribution to Gaia theory.

Jenny Powys-Lybbe was an English teacher at Dartington Hall, a magnificent stately home set in 1,200 acres of Devon countryside. Twenty-one years younger than Lovelock, she wrote plays, read philosophy, performed in amateur drama productions and later wrote a book on Shakespeare's *Richard III*. Her grandfather was an industrialist who had made enough of a fortune through his steel-coking plant to give a new Bentley car to each of his four daughters. It was his money that put Jenny through a private convent school and paid for her English and Drama degree at Bristol University. She married young, had a child and then divorced and drifted towards Dartington, where her mother bought her a six-bedroom house and her son, Christopher, could get a free education at the school where she worked.

Run as a trust on the progressive lines set down sixty years earlier by the pioneers of rural reconstruction, Leonard and Dorothy Elmhirst, the Dartington estate had become a hothouse of art, spiritualism and ecological thinking and a spawning ground for some of Britain's leading environmental thinkers. In 1982, the chairman of the Dartington Trust, John Lane, asked Jenny to invite Lovelock to give a workshop, a concept that was then so new that she had to explain it to him. He was all too keen to participate, not least because the attractive young woman offered to put him up in her home during his stay. They became lovers, and his well-attended talk, which was to be the first of many at Dartington, marked the start of Lovelock's transformation from 'bought man of industry' to environmental guru.

This was no mere PR makeover. Lovelock felt he was at the end of his life and had no cause or inclination to manipulate his reputation. He was still first and foremost a man of science, but he was open to new ways of thinking. 'Dartington was the "in" place to be. He wanted to be part of that set,' Powys-Lybbe remembered. 'Everybody seduced everybody . . . My first impression of Jim was that he was a very naughty man. He had only one thing on his mind. He was very good at it, and he liked to do it a lot. He had a healthy appetite.' Their relationship lasted for many years, always open and non-exclusive. 'There were others at Dartington,' Powys-Lybbe recalled. 'You could never rely on him staying out of bed. I was probably his favourite when he was here, but it could change.' Everyone there knew they were together. Lovelock would talk about Jenny to his children and once took her to Coombe Mill to meet his wife.

The affair was tough on Helen. She adored her husband and made sure he knew it. After one impassioned outburst by her, Jim wrote to tell his wife how much she meant to him, and he reminisced about the past: 'It moved me very much and took me back to our young days together in the war. How could you ever wonder why I married you? Don't you remember how good it was?'[7]

Lovelock would later blame this 'barmy' period in his life on the influence of opiate painkillers. But Dartington was an important waystation in his intellectual development. After being wounded by the 'bought man of industry' jibe from Commoner and others, he had scaled back his consultancies with chemical companies and the oil industry and started looking for new ideas. For the first time, he was mixing with environmental campaigners as friends rather than adversaries. In some of his public speeches over the following years, he even started to describe himself as a green.

Powys-Lybbe enjoyed his company, delighted in his sense of humour and felt privileged to be with someone she considered exceptional. Lovelock would join her at the Catholic Church on Sundays. They hiked together in Dartmoor. Her son Christopher, a talented computer programmer, worked as Lovelock's apprentice. And she accompanied him on his many trips to hospital during this period. The frequency of these visits increased alarmingly in 1982, when Lovelock underwent emergency open heart surgery and then suffered an infection that damaged his urethra so badly that he needed several operations to rebuild it. He felt the treatment was never-ending. 'Whenever I came out of hospital, I was pitchforked into something else. I was almost like a soldier on the battlefield.' His daughter Christine recalled the anxiety she felt at that time for both her parents: 'They were both battling for their lives. We all expected him to die.'

Lovelock was not the type to sink into depression. He kept himself busy, looking for solutions to other problems and trying his hand at science fiction, with two novels co-written with Michael Allaby titled *The Great Extinction* and *The Greening of Mars*. Their limitations were made clear in a hand-written note from Isaac Asimov, who observed: 'It isn't really easy to write science fiction and not everyone can do it.' Jim moved on.

Acknowledging his own fragility was difficult, but Jenny helped to change this. She encouraged Lovelock to face up to his mortality. Following a vogue at the time, she had become fascinated by *The Tibetan Book of the Dead* and started organising death workshops. 'I made my living out of them. That was my chief interest. We did many in London . . . It was for those who wanted a practice run, and many do,' she recalled.

Lovelock participated in one session arranged at an old Victorian house on Talgarth Road in Fulham. The amateur dramatist Powys-Lybbe told the small group to lie down on yoga mats in a darkened room, close their eyes and imagine they were walking through a flower-filled meadow down to the bank of a pellucid river. Once they crossed the small bridge to the other side, she told them they would reach an island, where death awaited them and they should open themselves to embrace it, conceptualising what form it might take in as much detail as possible. Coming so soon after his mother's death – and with his own heart problem and his wife's MS rapidly deteriorating – it felt all too easy for Lovelock to imagine. 'It was troubling stuff,' he recalled. 'My death was connected with the hydrogen bomb.

It looked like a barrel-shaped stove. But I've actually seen one and it is filled with what looks like silver golf balls.'

Lovelock later dismissed this as morbid nonsense but, after facing up to death, there was a nuanced shift in his public speeches about Gaia around this time. Out went the indestructible goddess which had delighted Shell executives because she seemed big enough and strong enough to absorb any industrial waste. And in came the image of a sick patient, a fragile Planet Earth that was vulnerable to human abuse and in need of planetary medicine. Forced to contemplate his own demise, Lovelock had started to consider that Gaia, too, might be mortal. From this point on, he often drew comparisons between himself and the Earth goddess.

More broadly, the meaning of Gaia was starting to slip out of his grasp, having been adopted and reinterpreted by a dizzily diverse following of cybernetic analysts, eco-modernists, diplomats, New Age gurus, witches and wizards. Spiritual groups were fixated by the suggestion of a grand design or providence. Lovelock – the avowed agnostic but ever curious spiritual explorer – constantly had to deny his theory had anything to do with creationism and to assert, instead, that the Earth system was the result of Darwinist evolution through random chance. Yet, the former Quaker was more than happy to seek common ground with religious supporters. He accepted an invitation to preach at the world's largest Anglican church, St John the Divine, in New York, corresponded enthusiastically with the Bishop of Birmingham and joined Lindisfarne Association gatherings to discuss the creation of a new planetary culture.

It was there he met the American poet Gary Snyder who was inspired to write a volume of poetry that ended with a piece entitled 'Gaia':

> Deep blue sea baby,
> Deep blue sea.
>   Ge, Gaia
> Seed syllable, 'ah!',[8]

In today's parlance, Gaia was now viral, a proto-meme that came to signify anything to anyone. It sparkled as a brand name in consumer culture, masked greenwashing campaigns for industry and declared a global, earthy perspective in art, engineering and alternative medicine. In the years to come, there were Gaia films, Gaia art galleries, Gaia rock groups, Gaia designer dolls, Gaia air conditioners, Gaia birthing centres, Gaia therapy collectives, Gaia healing sessions, Gaia spiritual workshops, Gaia yoga techniques, Gaia crystal scent drops, Gaia chocolates, Gaia tea, Gaia coffee, Gaia beer, Gaia whisky, Gaia fast-food outlets, Gaia nail-care kits, Gaia fashion chains, Gaia necklaces, Gaia video games, Gaia skincare, Gaia premium toilet paper, Gaia organic air-freshener, Gaia eco-friendly cutlery, Gaia 'Afrodisiacs', Gaia biodegradable vibrators, Gaia petcare, Gaia corporate consultancies, Gaia minivans, Gaia condominiums, Gaia biotechnologies, a Gaia supercomputer owned by the Petrobras oil company, a Gaia sustainability training course for chemical engineers and even a Gaia steakhouse in New Orleans, Louisiana. Little of this had anything to do with Lovelock and Margulis's theory, other than the fact that they had reawakened the ancient idea of a living Earth and then boosted it with scientific steroids.

Academics could no longer ignore Gaia. Neo-Darwinist biologists had made careers out of ever narrower specialisations and an absolute faith in the idea of evolution as a survival of the fittest. Lovelock and Margulis posed a diametric challenge by looking at the entire world system in a holistic way that emphasised the interdependence of species and the idea that each individual life on Earth had agency.

For many scientists, their Earth Goddess theory sounded a lot like religious creationism, which they had been fighting against for years. Biologist Ford Doolittle was the first on the offensive with a 1981 article titled 'Is nature really motherly?' in which he argued Gaia theory was unscientific because it failed to explain how feedback mechanisms functioned. Another famous biologist, Richard Dawkins, said Lovelock and Margulis's theory contradicted Darwinist principles and devoted several pages of his 1982 bestseller *The Extended Phenotype* to the argument that organisms could not act in tandem because this would require planning and collaboration. Stephen Jay Gould later dismissed Gaia as 'a metaphor, not a mechanism'.[9] Many others, including Robert Trivers, Robert May, W.D. Hamilton and George Williams, expressed varying degrees of scepticism.

Margulis led the counter-attack. She condemned the neo-Darwinists as 'entrenched servants of greedy masters'.[10] Unlike Lovelock, she was conscious of the political and environmental undertones of this debate. Neo-Darwinism, she wrote later, was a 'preoccupation with the romantic Victorian conception of evolution as a prolonged and bloody battle' that would prevail until 'overpopulation (with its concomitant toxic water, polluted airways and garbage) destroys technological civilization and its money-making, stockpiling thought-collective of which neo-Darwinism is only a tiny part'.[11]

The prevailing ideology in the 1980s was the neo-conservatism of Margaret Thatcher and Ronald Reagan, which argued that economic competition was human nature. Far-right fascist groups, such as the National Front in Britain, went further still by adopting Dawkins' 'selfish gene' to support racist and eugenic views. By contrast, the interdependent world view of Gaia seemed to be closer to left-wing theories of societal integration, eco-feminism and environmental stewardship.

Lovelock was bruised by the intellectual battering of his peers, but it was not in his character to lash out or give up. He remained positive, confronted the problem head-on and looked for solutions. He respected the neo-Darwinists. In an unfailingly polite letter to Dawkins, Jim wrote: 'You have approached this topic from the ultra microscopic level where genes exist. I have done so by looking at the Earth as a planet from the outside. It hardly is surprising that we don't see eye to eye.' He wanted to persuade his critics, by providing proof of his theory on their terms. This resulted in a falling-out with Lynn, who was not convinced by the technology-centred, data-modelling approach of modern science. It also posed an enormous challenge to the 'arithmetically blind' Lovelock.

He got around this by buying one of the most advanced Hewlett-Packard computers of that time, learning BASIC programming code and taking inspiration from his multi-talented postgraduate assistant Andrew Watson, his son Andrew, who was a computer programmer, and Jenny's son, Christopher, who knew how to code. Lovelock and Watson built a computer model to demonstrate how a self-regulating Earth system could have emerged through natural selection. They called this parable Daisyworld.

This was not long after the age of Atari Space Invaders, so the graphics were rudimentary, but it was the concept that was important. The model used light and dark flowers for positive and negative feedbacks. The influential climate scientist Steve Schneider described best how this model simulated aeons of evolution:

> The sun heats up over hundreds of millions of years, black daisies approach their optimum temperatures, become more fit, and thereby increase their numbers, causing the albedo (reflectivity) of the planet's surface to drop. This is a positive feedback, because while more sunlight is absorbed by the dark flowers the planet further warms. Black daisies increase until the temperature passes their fitness peak and moves into the fitness range for the white daisies. These then begin to multiply and replace the black daisies; this shift increases the planet's albedo, which serves as a negative feedback on further warming. The planet's overall temperature is stabilised for aeons even though the sun inexorably increases its luminosity. But eventually, the white daisies are heated past their fitness range and can't resist further warming. The biota then collapses and temperatures rise rapidly to the level an inorganic rock would experience.[12]

Watson and Lovelock's paper on Daisyworld was published in 1983 in *Tellus*.[13] It has since been cited as the first theoretical model of how living beings interact with their global environment.[14] Watson felt it was a turning point in Lovelock's life: 'In many ways he had been going through a low period, but he was very pleased with that work. That simple model became incredibly influential. He was thinking very hard about Gaia at that time. He was thinking of the criticisms of it and how one might answer them. Jim has always been a sunny person.'

Gaia started to gain traction. Neo-Darwinist opporobrium had raised the profile of the theory. Along with the Dartington lectures, his book and the increasing public profile, the hypothesis had become one of the most fashionable topics in the Earth sciences. The BBC produced a deep-dive documentary into the subject. Gaia attracted the interest of ecological philosophers such as David Abram, who explored the proposition in new ways: 'I don't think Lovelock noticed the most radical thing about his work, which was the interiority of the external world... The idea of an organism implicates us as being inside.'

The theory was frequently discussed in *New Scientist* and *CoEvolution Quarterly*, and Lovelock felt it would be given more of a sympathetic hearing among American geologists and climatologists than British evolutionary biologists. Stephen Schneider was sceptical about Gaia, but he saw it as a useful stimulus for research into life's effects on the climate. Amherst College, Massachusetts, hosted a couple of related gatherings in the mid 1980s and then in 1987, at Schneider's prompting, the American Geophysical Union arranged a weeklong Chapman conference on Gaia in San Diego.

This spectacle drew more attention than any geophysical conference before or since. As the *San Francisco Examiner* reported:

> It was an epochal event – a veritable United Nations of scientists in one large room, debating a single idea: the Gaia Hypothesis... Microbiologists challenged

atmospheric scientists. Oceanographers listened intently to volcanologists. Population biologists argued with geologists. Meteorologists, marine biologists, geochemists, geophysicists, botanists, space physicists, exobiologists, mathematicians and computer scientists wrangled, guffawed, laughed, drank, ate, quibbled and quarrelled through five long days.[15]

The author Lawrence E. Joseph observed that there was also a cluster of New Age followers: 'What Gaian event would be complete without cameo appearances from the likes of Brother John from the Bay Area's Institute of Immortalism, to balance off all of the left-brain karma.'[16]

Despite the circus-like atmosphere, Gaia was being taken more seriously by more scientists than ever before. Lovelock began to feel he was once again in with a chance of securing the recognition he craved. But, just as had been the case in Stockholm with Barry Commoner a decade earlier, the rug was pulled from under him. This time by a young Berkeley geomorphologist named James Kirchner, who mocked Gaia in a brilliant but utterly derisory oration from the stage in the hotel ballroom on the evening of the third day. It was the 'most vicious' speech of the conference, according to observers.

In a lengthy and bruising critique, Kirchner challenged the very validity of Gaia as a topic for discussion in any of its five main guises. The put-downs were summarised by Joseph:

1) influential Gaia, the formulation that life is but one participant in the global system, was dismissed as old news

2) co-evolutionary Gaia, that life and environment evolve as a coupled system, was condemned as unoriginal and totally misleading

3) homeostatic Gaia, the traditional, strong Gaian notion of biotic control of the global environment, was shunted aside as ill-defined and circular

4) teleological Gaia, the early implication that Gaia operated with intent and purpose, was criticised as a 'transparent teleology'

5) optimising Gaia, with life's collective purpose specified as creating a perfect planet, was denounced a tautological and internally contradictory

The audience was stunned, first by the strength of the verbal assault and then by the weakness of Lovelock's response. Instead of a point-by-point rebuttal, Gaia's elder statesman raised his hand, mounted the stage, expressed his gratitude to the young man and made a feeble 90-second defence of his theory.

Witnesses could not believe what they were seeing. 'It was dramatic as hell. This guy Kirchner thought he was putting the nail in the coffin of Gaia,' recalled Abram, who was disappointed that his mentor failed to put up a fight. 'Jim conceded much too much. He said he no longer wanted to defend strong Gaia and resorted to a mechanistic argument for a weak Gaia. This was disturbing. For me, the main appeal of Gaia was that it got us out of a mechanistic way of thinking. Lynn did better, but Jim was so deflated that it took away my ability to defend them. It was traumatic.'

At the time, Lovelock tried to wave the incident away. The Chapman conference had been a success in many other ways, generating widespread appreciation of so-called weak Gaia – which was a metaphor more than a theory – as a useful tool to open up new questions about the Earth system. In this important regard, it was a turning point. But years later, Jim admitted he had felt crushed to hear the esteemed audience guffawing at his theory. The humiliation took him back to the put-downs he had suffered at Strand School, when he had coped by imagining himself as a stick insect that could disappear into the background: 'I was devastated. I thought all that work was a waste of time.'

A separate dispute with other US scientists saw Lovelock nudged out of the GAGE network, even though he had designed the equipment, identified the locations and set up the first monitoring station. His beloved Adrigole was closed down and his friend and neighbour Michael O'Sullivan put out of a job. It seemed nothing was going right.

As 1987 drew to a close, Jim was despondent. He later told journalists that he felt his work on Gaia had been a complete loss and he had wasted twenty years without getting anywhere.[17] Such doubts reflected torrid personal problems and poor health. But he never completely gave up on Gaia. Far from it. Something strange, almost magical, was happening. Now that he recognised the fragility of himself and the Earth, he came to appreciate the importance of depending on others. His feelings toward the goddess had changed. He believed she needed him and might repay his devotion. 'Be good to Gaia, and Gaia will be good to you' became his new mantra. Beyond the technobabble of cybernetic feedback loops, he had come to realise that Gaia was first and foremost about a relationship.

Sandra

'Love is a defiance of physical laws. The more love you have, the more you generate, the more you have and the more those around you have. It augments and then augments its augmentation.'

The Gaia statue in the Lovelock garden started out as a joke. That was probably inevitable in a bohemian family of irreverent mischief-makers. No first-time visitor to Coombe Mill, or Bowerchalke before it, could be left in any doubt that they were in the presence of unapologetic eccentrics. The family kept peacocks on the lawn and goldfish in the swimming pool. One visitor was surprised to find Andrew's girlfriend sunbathing naked. Christine, now one of the best women runners in Britain, was a whirr of energy. John, meanwhile, was still chasing storms. And Jane? Well, she was the exception that proved the rule. As the most grounded and sociable of her siblings, she had left home as soon as she could, marrying an Irishman and raising a family in Japan, France, Saudi Arabia and Holland. Their mother, Helen, meanwhile, was the gatekeeper and account manager, who, notwithstanding her MS, would rustle up food, crack open wine, regale visitors with dirty jokes, handle mail, manage the money, filter out 'sycophants' and keep everything and everyone on an even keel.

The statue of a buxom, long-haired woman became part of this realm a year or two after Lovelock decided to call his new theory after the Greek goddess Gaia. When the family heard the name, they were typically irreverent. Andrew immediately quipped: 'Can I have the rights to print the T-shirts?' and then sarcastically called his dad 'guru' for the next few months. There was no ill intent. Nobody in the family was allowed to take themselves too seriously. That was one of Helen's precepts. It kept people, especially Jim, down to earth. The statue was adopted in the same spirit.

The Lovelocks first saw the figure in 1972. Jim, Helen, Christine and Andrew had made a weekday excursion to Cranborne Manor Garden Centre, an emporium of plants, pots, seeds, ceramic gnomes and other accessories that was a twenty-minute drive from Bowerchalke and a favourite spot for green-fingered Helen. As they cruised the tree-lined avenue that led up to the stately home, they noticed a paddock full of forlorn-seeming concrete replica statues. 'I think they were part of a sale, unwanted; it was almost like an Orson Welles story, we felt sad for them, created to live in the grounds of stately homes but unwanted in a world that was even then becoming more like suburbia, with stately homes being taxed out of existence,' Christine recalled. 'I or Andrew said: "Why don't we buy that one with the grapes and call her Gaia?" It would have tickled Mum and Dad's sense of humour, pretending that we thought we looked grand with a big statue on our small top lawn. We all laughed. It was a sort of a joke to begin with. It was ostentatious and we weren't like that.'

The stone image was winsome, even wanton, a young woman with her robes and hair in disarray, bearing grapes – an allusion to wine. The original sculptor was probably aiming for a Bacchanalian or Saturnian nymph to lighten up the garden of an English gentleman. Once the Lovelocks got hold of her, though, she was upgraded to a goddess and given a prime spot on the upper lawn in Bowerchalke.

Although the statue started as a figure of fun, she became part of Lovelock's life. When the family moved to Coombe Mill, there was no doubt that the statue must go with them.

Gaia was placed close to the house so that she was one of the first things visitors would see if they were brave enough to drive beyond the radiation warning sign at the gate. Lovelock planted trees in the background, which made the figure irresistible to the innumerable magazine and newspaper photographers who came looking for an iconic image that could represent his hypothesis about nature. He would spin stories (a lifetime source of pleasure) around the statue, saying he had it commissioned. Over time, the concrete image appeared in so many articles and book covers that she became, at least in the public mind, the face of Gaia.

To Lovelock, she was something more again. As a former Quaker who had flirted at various times with Catholicism and Anglicanism, he was an open-minded non-believer who had always been fascinated by religion. It was only a matter of time, therefore, that he began to look at Gaia – both the myth and the statue – in a way that went far beyond the original concept of a homeostatic biogeochemical feedback system.

In the 1970s and '80s, he suffered for Gaia and with Gaia. They went through so much together that he felt a sense of solidarity. There was also a sense of gratitude, bordering on reverence for the natural world that she represented. He would talk and even pray to the statue, which he started to call Ma Ga (Mother Gaia). This had to be done in private because he knew critics would jump on this as evidence that his theory truly was creationist and teleological. It was not that simple. For Jim, Ma Ga was both 'Mother Earth' and a 'Tough Bitch', powerful but vulnerable, not unlike Nellie.

In short, by April 1988 his relationship with Gaia had gone far beyond what might be expected of an agnostic scientist. In that giddy year, it would hit a whole new level of intimacy with the appearance in his life of a woman he came to see as the living embodiment of the goddess. Nothing would ever be the same again.

How to measure the rapture of a man in love? By the giddy prose of the letters to his beloved? By his ability to defy the years with astonishing acts of sexual prowess? By personal proclamations of rebirth and new life? By distressing displays of indifference to those who had previously been at the centre of his life? Or by an overbrimming confidence in his power to set the world to rights? Take your pick. It does not matter. By any or all of these counts, James Ephraim Lovelock found himself, at the age of sixty-nine, in a state of ecstatic, romantic delirium.

For the first time in more than a decade, he suddenly felt life coursing through his veins. Death was vanquished. Anything seemed possible. Everything was full of meaning and joy. He was quoting poetry, immersing himself in opera, dancing to pop music, caressing the breasts of the deity in his garden and mouthing prayers of thanks to Mother Earth. There was no doubt in his mind. No fear. No remorse. Suddenly, the way ahead was clear. After a lifelong search, he believed he had found what he was looking for.

Sandra Orchard was an elegant, well-read American woman who was nineteen years younger than Lovelock. She was an unlikely goddess. After growing up in St Louis, Missouri, she had worked as an English literature teacher and a commercial voice-over artist, divorced a body-builder in Florida, remarried an opera-loving flower arranger named David Orchard and reinvented herself in London as an event organiser working for Wilfred Grenfell, an Eton and Oxford educated man whose sister, the status-conscious Orchard was proud to observe, was married to the Duke of Richmond. Compared to Hitchcock or Margulis, she had never been a leader in any academic field. Unlike Helen, she had no experience of mothering. But she knew how to compose a gathering and put life into a party. Cheerful and easy-going, her nature could not have been further from that of Nellie, let alone the mythical tough bitch who had punished her lover with castration. But she did share Lovelock's wicked sense of humour, curiosity about the world and passion for the outdoors. Sandy — as he came to know her — seemed a second chance at life. Jim would reinvent his world for her.

They met at the Global Forum of Spiritual and Parliamentary Leaders on Human Survival, an event that exemplified the late 1980s confidence in humanity's ability to come together and make a better world. With Mikhail Gorbachev in power in the Soviet Union, the Cold War was starting to thaw. In South Africa, apartheid was in its death throes and Nelson Mandela would soon be free. The war in Northern Ireland had long since passed its bloodiest peak and the different sides would sign the Good Friday agreement in the coming decade. All the world's governments had recently signed what is still the planet's most successful environmental agreement, the Montreal Protocol, which phased out CFCs and other ozone-depleting substances that Jim had helped to identify the previous decade. At the opening of the forum, it seemed to many people that the planet and its dominant species were in a healing mode. In retrospect, it was probably more telling that the sponsor of this great spiritual event was an oil and gas company, the US-based Energy Development Corporation.

Sandy had arranged the invitations, logistics and schedule of the event, which brought together 140 of the most influential men and women from sixty nations for five days of talks in Oxford. Among the dignitaries were representatives of Hinduism, Buddhism, Christianity, Islam, Judaism, Jainism, Sikhism and Shinto as well as African and American indigenous leaders. The Tibetan god-king the Dalai Lama was there, as was the Catholic champion of the poor, Mother Theresa; the Anglican Archbishop of Canterbury, Robert Runcie; and the founder of the tree-planting Green Belt Movement in Africa, Wangari Maathai. Standing beneath a 15-foot photo of the Earth which was to symbolically preside over the deliberations in Oxford's Old Town Hall, Carl Sagan described the event as 'the meeting of the cousins, the gathering of the human tribes'.[1]

When it was Lovelock's turn to speak, he set out his Gaia theory in the greenest, most spiritual and emotional terms yet. Harking back to Isaac Newton and other natural philosophers of the past, he reminded his audience that the life of a scientist had once been both 'deeply sensuous and deeply religious'. For the first time in his speeches, he talked of the importance of love: 'Curiosity is the principal motivation of the natural philosopher and curiosity also is an intimate part of the process of loving. Being curious about and getting to know a person or the natural world leads to a loving relationship.' This was revelatory, tying up many threads of his life, combining his quests for knowledge and love, and articulating for the first time what drove him.

The choice of words was serendipitous. The previous morning, he had briefly met the 'stunningly attractive' woman who would become his second wife in the quadrangle of Christ Church College. She was among the organisers who were checking in the guests and handing them name badges and programmes. Sandy was charmed by this humble man with tousled white hair and thick, black-framed glasses: 'I hadn't heard of him before. Isn't that awful,' she laughed. 'We started talking. He was so modest. He was wonderful.'

They agreed to meet for lunch, a date that fell through, and then at the official concert the following evening, where they sat uncomfortably side by side and discovered a shared antipathy for country and western music. It was not until the third night that their stars aligned. The setting was a grand dinner in the library of Blenheim Palace, where the champagne flowed freely and guests were serenaded by a string quartet and a South Korean violinist. Jim and Sandy were frustratingly seated too far apart to talk so it was not until they headed back to the coaches at the end of the evening that they could meet. Pleasantly stimulated by smart conversation and fine wine, they caught sight of one another in the corridor and fell into a passionate embrace and then sat together on the coach holding hands, saying little, until they reached the Randolph Hotel.

'For some reason it felt the most natural thing in the world to be in each other's arms,' Sandy recalled. 'It was a whole night of sex interspersed with limericks,' Lovelock laughed, making his wife giggle as he recited one of them:

> There was a young fellow from Pons,
> Who was fined for bothering swans.
> Helpful, the porter,
> Said, 'please take my daughter,
> The swans are reserved for the dons.'

Lovelock was sixty-nine years old, in poor health and, by his own account, had endured 'enforced celibacy' for several years. That night, he felt reborn. 'I've never been one to feel sorry for myself, but when I got together with Sandy, I was ecstatic. It's not a bad thing to have that joy postponed.' They did not waste a single minute on sleep. The next morning, they shared breakfast at the Randolph, watching the world pass by in the rain outside, pitying anyone who could not experience their joy, and worrying what this might mean for their marriages.

After a night like that, they could not bear to be apart. Lovelock extended his stay in Oxford, then arranged trysts in a flat in Notting Hill, then a trip to Fallujah. He introduced Sandy to an ever wider circle of his friends. For him, the relationship was a second chance: 'I don't think it's just because I had a heart attack but there was a point when I really felt I had no future. Gaia and Sandy helped me enormously.'

In his love letters to Sandy at this time, Lovelock reveals that he prayed about her to Ma Ga (Mother Gaia): 'Having spoken to our lady (Ma Ga) this morning your phone call was not unexpected but all that I had asked for was that you called so that I could hear your gentle voice again.'[2] In his besotted mind, the goddess and his lover began to merge. In another letter Jim wrote of his 'malignant longing' for Sandy while she was away in southern Europe: 'Now I must go and have a laugh with

Ma Ga and cup her breasts in my hands. If you are the embodiment of her, then maybe she is that of you and you will feel it in Spain . . . I shall never know unless I try.'[3]

Lovelock felt a need to share his joy with everyone. He confided in his daughter Christine, who was uneasy about what the affair would mean for her mother, but also glad to see her father looking happier and healthier than he had been for many years. 'Sandy blew his mind. It was like he'd got a new life,' she said. 'It was so good to see Dad so happy. He had been through such trauma.' His friends in the secret service sent him a framed photo of the multitude of attendees at the spiritual conference with a compliments slip written in green ink (the colour of messages from 'Command', the head of the secret services), that read: 'Blenheim – Sandy "a gift from Gaia".'

The most eloquent letter of support for his new relationship was inevitably from Lynn Margulis and her partner, Ricardo Guerrero, in Barcelona:

22 May 1988

What a joy to see you happy and fulfilled. Although I never thought about it, if you had asked me, I would have denied that you were still to find the love of your life. Last time with Jenny (of Exeter) you fretted far more than you rejoiced. With Sandy, you don't fret at all, you seem all joy. That fact is a delight to your old friends who really love you, who hate to see you suffer the repression, suppression and depression that is the normal response to an increasingly ill wife and the weight of people's dependencies. I don't believe in conservation of love. Love is a defiance of physical laws. The more love you have, the more you generate, the more you have and the more those around you have. It augments and then augments its augmentation. Ricardo and I – except for earlier formal meetings of no importance – made love on the night of our first meeting. Never before had we spent three days in bed – I missed an aeroplane! Here we are still generating the same emotional-intellectual-physical passion five years later. I believe in the definitive unequivocal lightning bolts of love – but I've seen many people who are not fortunate enough to be struck. One must love oneself before anything is left over for one's children, lovers, friends et cetera. But there is no conservation . . . To reiterate: we love you dearly, we'll help you in any way we can. The two of you make us very happy. It is never too late.

Please keep in touch.

Lynn

PS you will be a far finer husband, father and grandfather. This is what I mean about 'lack of conservation' of love. I'm a great admirer of your unfettered, uncomplicated honesty and affectionate demeanour and always knew you had it in you. I was only worried about the nearly permanent lack of opportunity.

Many things seemed to click for Lovelock around this time. Gaia was being taken seriously. Although academics still bristled at the name, former critics, such as the biologists Bill Hamilton and John Maynard Smith, were starting to come round. A study that Lovelock had co-published the previous year, proposing a negative feedback loop between ocean ecosystems and the Earth's climate, was well received. This important paper argued that phytoplankton produce dimethyl sulphide, which is responsive to variations in climate forcing and this helps to regulate the temperature of the Earth's atmosphere. It was pure Gaia, although to get around the critics at *Nature*, they didn't use the 'G-word'. Instead, the hypothesis was named after the initials of the four co-authors, and the duly published CLAW hypothesis received considerable acclaim.[4]

Similarly, Lovelock's second book, *The Ages of Gaia*, which had been edited this time by Margulis, was acknowledged by academics as superior to its predecessor. Described as a 'biography of the Earth', it combined the insights of Daisyworld with a palaeoclimatological history of the planet and observations on the response of the faith community to Gaia. Here was Jim in full planetary doctor mode, arguing that the health of the planetary organism had to come ahead of the well-being of any of its individual species, including humanity. This time, he very specifically refuted any suggestion that *Gaia* 'gives industry the green light to pollute at will'. Years later, this book would be considered his masterpiece.

The publishers, W.W. Norton, and sponsors, the Commonwealth Fund, arranged a gala party for the launch, which Jim and Sandy flew to New York to attend. This indirectly led to ructions at home. Lovelock had tried to keep his affair a secret from his sick wife: 'I couldn't tell Helen. She was in a bad state already.' But as with his previous affairs, he was incompetent at concealing his feelings. Under Sandy's influence, his life was suddenly filled with music: Mozart, Bartok and Prokofiev. In a letter to his lover, he sees the two of them creating a new 'domain' together:

2 July 1988

I'm listening to the first act of Don Giovanni to be in tune with you, my gorgeous mentor . . . it is the way of Gaia to persist in stability and resist perturbations, until suddenly when the stress has become unbearable, she must jump to a new stable state or die. For this domain, here at Coombe Mill, the stresses are already near intolerable.

Over the following months, those stresses at Coombe Mill reached breaking point when the publisher of the second Gaia book, Ed Barber, sent a letter to Lovelock about the launch party, saying how much he enjoyed meeting 'your lady'. He could not have known that all of Jim's mail was opened by his wife. Helen was devastated, and Jim tried to deny his infidelity, but his excuses were unconvincing.

He made things worse by travelling with Sandy to Perugia, Italy, for a seven-day Lindisfarne Association meeting, and then

Sandra

insisting Sandy be made part of the Gaia Foundation charity he was in the process of setting up. Lovelock once again proved that a scientific genius can also be an empathetic dunce. He was focusing on Sandy to the detriment of all else. In September, he wrote to his lover about the difficulty of being together: 'I don't know yet how, but Ma Ga will solve it with a little help from us as she always does.'

In December, he wrote to his beloved with another saucy limerick and an entreaty that the couple devote themselves to Gaia and each other in the coming twelve months:

There was a young girl of Pitlochry
Had sex with a chap in a rockery
She said to her chum
The stones hurt my bum
It isn't a fuck, it's a mockery.
Let's make 1989 Gaia's year and look after our domain and fill it with laughter and lingering lust.

The discovery of another infidelity was the beginning of the end for Helen. 'It was almost certainly the stress of his affair that made her ill at the end. Stress is very bad for MS sufferers,' Christine surmised. 'Maybe that is why she went downhill so quickly.' The speed of the decline shocked her doctor, who sent her to a hospital in Plymouth on 23 January 1989 for what he assumed would be a stay for just a day or two. Helen died two weeks later surrounded by family and friends. The death certificate notes that she died of bronchial pneumonia and multiple sclerosis. Her body was laid to rest in the grounds of Coombe Mill.

The children were devastated. Helen had been the pillar of the family, making sure everyone was looked after while their father travelled for work. She was far more grounded than Jim, but also fun to be around. In letters, she would cheerfully exhort her daughter, Christine with lines such as 'chin up, chest out and be bossy' and mischievously sign off with 'your evil old woman, satan's wife'. Neighbours wondered how Lovelock would cope without a woman they referred to as 'his backbone'.

With the exception of Christine, none of the children knew about Sandy until the day their mother died, when she turned up at the hospital to support Jim. This was a shock. Jane observed that her father and his partner seemed 'too happy'. Within two weeks, Sandy had installed herself in the house and started working as a personal assistant to Lovelock. 'Sandy took over Mum's role seamlessly,' Christine recalled. 'It was difficult and uncomfortable. It had been a family home and there was Sandy packing up all of Mum's stuff. We thought it should be us . . . It was too much. She and Dad were like babes in the wood. They didn't get it.'

Social conventions were then completely discarded when Sandy's husband, David Orchard, moved into Coombe Mill to complete a very open relationship triangle. 'It became so weird,' Christine recalled. 'There was a party for Dad's birthday in the pub and everyone said: "Who is that with him?"' The secret services were also suspicious and insisted on doing a background check on David.

Lovelock had broached the idea of them all being friends several months earlier: 'Dear David, I think that this is the most difficult letter I have ever had to write . . . Do you think that we can all manage some kind of emotional accommodation? I sincerely hope so.'[5] This proposal was not as outlandish as it might sound. Sandy said her marriage to David had already turned into a sexless friendship. After a lifetime of smoking he had just been diagnosed with inoperable lung cancer. He decided he would rather spend the end of his life at Coombe Mill than with his sister and her family in a council estate in Pimlico.

The arrangement seems to have been a success. The threesome went on holiday to Sicily, hiking on Dartmoor and on weekend visits to country homes. 'Strangely, it worked like a dream,' Lovelock remembered. 'It was like marrying a widow whose husband was still alive. He either liked me or tolerated me so much that I didn't notice.' David died at Coombe Mill a year after arriving. Lovelock and Sandy married immediately afterwards – a very quiet registry ceremony followed by a meal at a Thai restaurant. To honour David's memory, they placed a bench engraved with his name on a hill at the farm.

At seventy years of age, Lovelock now ceased most of his scientific research. It was as if exploration were no longer necessary because he had found what he was looking for. He gave up his big laboratory in the house so that Sandy could use it as a library for her books. He continued to do the hands-on work required by the intelligence services in the barn, but there was to be little more pure science.

The consultancies for Shell and the chemical companies were fading. But there were other sources of money: donations to the Gaia charity, prize money and interest from previous contracts, including a lucrative commission for a massive perfluorocarbon tracer test for the US National Oceanic and Atmospheric Administration on the likely spread of fallout from a nuclear bomb, industrial accident or biological weapons attack on San Francisco.

The focus of his life was no longer experimentation, it was revelation. He wrote books, gave speeches, helped create the SimEarth video game and campaigned for his theory and the Earth. 'Gaia has become almost a way of life now and leaves little time for anything else,' he wrote to a friend.[6]

Improving Coombe Mill became an obsession. Lovelock planted thousands of trees, aiming to recreate the kind of English country landscape he and his father had happily wandered through in the 1930s. Ownership of the property was transferred to the Gaia charity. Jim reclassified the farm as a forestry area, thereby deliberately devaluing his home so there would be less incentive for future developers to clear the land for construction. In speeches and magazine articles, he increasingly stressed the importance of protecting the Amazon and other tropical rainforests, whose clearance he described as 'a vast, urgent and certain danger'.[7]

Guests to the house were carefully vetted by Sandy. The couple repeatedly changed the Coombe Mill telephone number to avoid 'pests, the intrusive media people and other nuisances'. When Margulis complained that she was also being cut off, her long-time collaborator was soothingly reassuring: 'Dear Lynn, please don't assume that my desperate attempts at a quiet life are meant to exclude you.'[8] But even Lovelock's children struggled to secure an audience.

Lovelock's circle of friends was changing. The petrochemical executives and PR men were kept at arm's length. In came environmentally conscious intellectuals, many of whom he had first met in Dartington, such as the editor of *The Ecologist*, Edward Goldsmith; the founder of the Ecology Party (now the Green Party), Jonathon Porritt; the deep ecologist Stephan Harding; and the Earl of Portland, Henry Bentinck. A frequent visitor and fellow hiker was Satish Kumar, a Jaian monk who was editor-in-chief of *Resurgence* magazine.

Kumar's first impression captures the Lovelock of that period: 'He was very shy and self-effacing, but quite passionate once we started talking. He said: "Everyone talks about human rights, but I want to talk about rights of nature, insects, birds, wasps, butterflies and other-than-humans. I want to stand up for Gaia."' Kumar invited him to give a series of inaugural lectures at Schumacher College, which he and Harding had founded in 1990 as a scientific-education offshoot of the Dartington Trust.

Lovelock's classes set the tone of the college for decades to come, influencing several generations of students, many of whom went on to become prominent figures in academia, climate research, politics and government. For Harding, an Oxford-trained zoologist, Lovelock's lectures – on planetary atmospheres, Mars landers, Gaia and Daisyworld – were a life-changing experience. 'I had been stuck in conventional science. After I met Jim, I realised my job was to teach Gaia,' he remembered. 'Gaia is the most holistic idea in the whole discipline of ecology.'

A politically important friendship developed at this time with Crispin Tickell, the government's top diplomat who was the permanent representative for the UK on the United Nations Security Council. He was a passionate environmentalist who had taken a year off earlier in his career to write the prescient 1977 book *Climatic Change and World Affairs*, which warned of the dangers of fossil-fuel emissions and the perils of thinking capitalist growth was an end in itself. He found a kindred spirit in Lovelock's first book and wrote to congratulate the author. They met in person at the launch of the second Gaia book in New York, which was the start of a close relationship that lasted until the end of their lives and helped Lovelock and his theory to reach a global audience.

Tickell was the most influential friend Lovelock had known since Rothschild. He served in the inner circle of British prime minister Margaret Thatcher, where his environmental views were always in a minority. 'I think my interest in climate change was always considered quite eccentric,' he remembered. Lovelock became the tip of his campaigning spear.

The statesman arranged for the scientist to attend a dinner at 10 Downing Street with several senior bureaucrats and captains of industry. Lovelock said he initially assumed he was just a place filler, until 'immediately after dinner, we were led into a big room. I had my back to the wall, but Thatcher made a beeline for me and talked animatedly about my book for twenty minutes.' Having been a chemistry student at Oxford before she entered politics, the premier knew what questions to ask and sought precision on what levels of carbon dioxide emissions would lead to dangerous levels of global warming. Was it really an urgent concern? Lovelock said his answer left the prime minister in no doubt that something needed to be done on the climate: 'It is important . . . This whole affair, it's a bit like the Falklands, it needs a strong leader to get us right.'

After this, Lovelock sent Thatcher a copy of two of his speeches. They had to be vetted by her private secretary, Charles Powell, who gave his enthusiastic approval. At the top of a copy of Lovelock's lecture to the Schumacher Society on 5 November 1988, he wrote: 'Prime Minister, An absolutely fascinating paper, which I do strongly recommend if you have the time.' He was still more enthusiastic about the address Jim made to the Friends of the Earth on 22 September 1989: 'Prime Minister, I think you will enjoy this.'

His boss paid sedulous attention to both, according to those familiar with her system of annotation. None of the text was marked with wiggly lines, which would have been a sure sign of disapproval. Many passages were highlighted and underlined, which generally meant she agreed. And there were multiple sections where Thatcher scrawled an arrow pointing left in the margin, which indicated an interesting point she wanted to discuss further. 'It is unusual to find more than a couple of these in a document, so her interest in the September 1989 lecture was of some intensity,' Thatcher Foundation archivist Chris Collins said.

The underlined parts included Lovelock's assertion that 'science is badly in need of greening', that 'environmental disorder' posed as great a threat as Adolf Hitler, and that Britain was a world leader in the search for solutions.

Among the passages that merited a left-pointing arrow in the margin was Lovelock's grim warning: 'What we are doing now to the world, by degrading land surfaces and by adding greenhouse gases to the air at an unprecedented rate, is new in the experience of the Earth. The consequences are unpredictable except to say that they will surely include surprises, some of which may be terrifying in human terms.'[9]

The prime minister seemed most enchanted, however, by Jim's Gaian vision of the Earth as a living system:

Sandra

If you find it hard to envisage our planet, mostly molten and incandescent rock, as alive, think of one of those giant redwood trees that grow on the west coast of the USA. They are alive but some are 3,000 years old. They are spires of lignin and cellulose weighing over 2,000 tones and 97% of them is dead. The wood inside and the bark outside support and protect a thin skin of living tissue of cells around the circumference of the tree. Just like the thin skin of living organisms around the circumference of the Earth.[10]

In the margin of this passage, which is highlighted and underlined in multiple areas, Thatcher scrawled: 'Marvellous.'

The head of the National Rivers Agency later wrote to tell Lovelock that when he and some colleagues went to see the prime minister, 'she had one scientific paper with her. The author was yourself. Clearly the good lady reads the right literature.' A delighted Lovelock replied: 'Thank you for telling me about the PM and Gaia. They are both very tough ladies and go together.'

Tickell was the main driver of the environmental agenda inside the government. At his encouragement, Thatcher organised a climate seminar at Downing Street with the aim of persuading her sceptical cabinet colleagues that there was no time to lose. Lovelock was invited, along with the future head of the United Nations IPCC, Robert Watson; the chief executive of the Met Office, Sir John Houghton; and another of Lovelock's old friends, Stephen Schneider, the Professor of Environmental Biology and Global Change at Stanford University who had organised the Chapman Conference on Gaia in San Diego. At a dinner that evening, leading environmentalists were called in along with senior members of government. Despite the hesitations of Chancellor Nigel Lawson, Thatcher convinced enough colleagues to take bolder steps. For a brief moment in history – the 'Gaia year' of 1989, as Jim presciently called it – the Iron Lady became an unlikely global champion of the environment.

In the darkling hours of a blustery winter's day, Sandy looked up from her correspondence as the Hewlett-Packard fax machine on the desk whirred into life. The sight of a crown, lion and unicorn emerged from the machine, followed by a capped letter header, 10 DOWNING STREET. She was excited. As soon as all the pages were printed, she took them to Jim and told him, with evident pride, that the prime minister wanted him to check the climate speech that she would give at the United Nations. Thatcher's private secretary, Powell, specifically asked: 'There are two or three points in the speech where she [the prime minister] would like to use entire sentences taken verbatim from your recent lectures. Would you be prepared for her to do this?'[11]

As he read through the lengthy script, Lovelock recognised data, references and phrases that had been lifted from the texts he had sent her. He scribbled a few suggestions and commended the thrust and ambition of the text. A few days later, Thatcher astonished the world with the most powerful environmental speech ever given by a global leader in such an exalted forum. The fact that it came from the mouth of a champion of deregulated markets and industrial expansion, made it all the more remarkable.

'Of all the challenges faced by the world community . . . one has grown clearer than any other in both urgency and importance — I refer to the threat to our global environment,' the speech began. 'What we are now doing to the world, by degrading the land surfaces, by polluting the waters and by adding greenhouse gases to the air at an unprecedented rate — all this is new in the experience of the Earth.' After this echo of Lovelock, she went on to propose solutions that must have sounded heretical to her free-market supporters. The champion of individualism, privatisation, competition and a small state said the international community needed to come together and strengthen global environmental governance. The proudly parsimonious housewife said it was time to spend money on the environment. She announced the UK would invest in a new centre of climate research, would commit an extra £100 million to tropical forestry and had recently doubled its contributions to the United Nations Environment Programme. She urged other governments to step up: 'We have to look forward not backward and we shall only succeed in dealing with the problems through a vast international, co-operative effort.'

The speech – and another she gave the same year at the Royal Society – had the fingerprints of Tickell all over it, echoed several Lovelockian turns of phrase and was infused with the spirit of Gaia.[12] Lovelock was thrilled. The following year, the Met Office opened the world's most advanced climate research institute, the Hadley Centre. In 1992, world leaders gathered for an Earth Summit in Rio de Janeiro where they established the framework conventions on climate change and biodiversity that are now the basis of annual COP meetings.

Jonathon Porritt, who was a critic of Thatcher on most other issues, said no politician before or since had ever come out so strongly for the environment. 'Those two speeches are beyond compare for a serving politician,' he remembered. 'It was a genuinely big movement, leading to 1992. If only world leaders had acted on insights being marshalled at that time.' The deregulation wing of the Conservative Party made sure that would not happen. A year later, Thatcher was gone, driven out of Downing Street by her own ministers over Europe, the poll tax and whisperings that she had started to put her global legacy above national and market interests. She would never again be such a strong advocate of action on the environment.[13]

Before departing, she rewarded Lovelock for his contribution to her speech and other work he had done on her behalf by signing up as a supporter of his Gaia charity and proposing him a Commander of the Order of the British Empire in her final New Year's honours list. At the award ceremony in Buckingham Palace, the Queen asked her usual question, 'What work do you do?' To which the seventy-year-old MI5 scientist prompted a royal chuckle with an exaggerated hush-hush gesture and the reply, 'I'm sorry I can't talk about it.' He later wrote to one associate that the CBE 'gives credibility both to Gaia and to me'. On a lighter note, he told another friend that for 'the next step I hope to be Lord Luvaduck of Broadwoodwidger', which was the name of a place near Coombe Mill that had tickled his sense of humour. Sandy dreamed that he would one day be given a knighthood so she could style herself 'Lady'.

Lovelock was now devoting his time to Gaia and the secret services. Essentially, this meant he was now dedicated not just to Sandy, but to his 'two great loves: the natural world and bombs'. He was closer to the establishment than ever before; invited to dinner at Buckingham Palace and several gatherings of spiritual leaders at St George's House in the grounds of Windsor Castle, hosted by the Duke of Edinburgh.

Meanwhile, he and Margulis led a series of international Gaia conferences in Oxford with funding by Norwegian shipping magnate Knut Kloster and organisational support from Tickell, who was by then Warden of Green College and President of the Royal Geographical Society. Participants lauded the interdisciplinary nature of the gatherings. 'These conferences were some of the most creative meetings I've been to,' Andrew Watson recalled. 'They were important in helping to gain acceptance for Gaia in biology and climate science. The mix of invitees included a number of "critical friends" who, while sceptical of Gaia, were prepared to engage in discussion, such as John Maynard Smith. From that time on, the evolutionary biologists began to think more creatively about non-neo-Darwinist models of evolution.'

In 1991, Lovelock published *Gaia: The Practical Science of Planetary Medicine*, extolling the importance of geophysiology – the neologism that had come to the fore in the wake of the Chapman conference. From 1990 onwards, he was a magnet for accolades and his opinion was sought by powerful statesmen. He won the Amsterdam Prize for the Environment. The President of the Czech Republic, Václav Havel, eulogised Gaia in his Liberty Medal acceptance oration in the United States. Japanese finance minister Kiichi Miyazawa asked his advice on whether Japan should contribute billions of dollars to the United States for a Superconducting Super Collider (Jim said he replied that this would be a waste of money because Europe was already working on a similar project).

In 1997, he was awarded the Blue Planet Prize, which was sometimes described as a Nobel for the Environment. Japan was a regular destination in this decade thanks to the hospitality of Mitsubishi, which paid Lovelock for consultations on its extensive nuclear business. Later, he was also presented with the highest accolade in the field of geology, the Wollaston Medal, in recognition of his influence on the development of Earth System Science.[14] In his acceptance speech, Jim memorably defined himself as a person who could 'bring separated things and ideas together and make the whole more than the sum of the parts.'

On his eightieth birthday, just before the turn of the millennium, life was going better than he could possibly have imagined a decade earlier. Many of the key concepts of Gaia were now widely accepted by Earth system scientists. Lovelock felt confident enough to publicly suggest that the Gaia theory could serve as a substitute for religion among agnostics. Unbeknownst to the outside world, he and Sandy were practising this themselves. They would sometimes spend time with the statue, using it as a focus for thoughts about Gaia and each other: 'If you put trust in Gaia, it can be a commitment as strong and as joyful as that of a good marriage.'[15]

To mark the turn of the millennium, Lovelock published a memoir titled *Homage to Gaia*. The couple gave active thanks for their blessings with a 'pilgrimage' around the southwest coastline of Britain. They divided the 630-mile trail from Poole via Land's End to Minehead into sections, staying at B&Bs along the way. It took the best part of the year – a remarkable achievement for a man of eighty-one and a woman of sixty-two. At the finish, they celebrated with a bottle of champagne. 'The coastal walk was one of the best things we did together,' Sandy remembered fondly.

Lovelock celebrated New Year 2000 in a state of profound personal contentment, though he foresaw calamity for the wider world. Thirty-three years earlier, his end-of-millennium prediction for Rothschild had been broadly right about environmental threats but completely wrong about the political response. Lovelock had assumed humanity would by now have been wise enough to replace fossil fuels with climate-friendly alternatives. But the reality, as Rothschild and others had demonstrated, was that oil firms resisted any attempt to curtail their lucrative businesses. More carbon dioxide was entering the atmosphere than ever before, disrupting Gaia's ability to regulate temperature. Lovelock was too old to worry for himself, but he feared for his grandchildren and other species. Though he was woefully ill-equipped to enter the world of politics, he felt it was time to speak up.

Nigel and Bruno

'They look to me as a guru and ask advice on saving the world. I always have to tell them I'm not sure I can help.'

In appearance and speech, Lovelock made an unlikely prophet of doom. Compared to the Old Testament image of a fiery and thunderous orator, he was humble, soft-spoken, self-effacing and generally preferred to write rather than speak. But whatever apocalyptic qualities he lacked in presentation, he more than made up for in the choice of his words and the sadly matter-of-fact tone of his horrifying prognostications:

> Few, even among climate scientists and ecologists, seem yet to realise fully the potential severity, or the imminence, of catastrophic global disaster . . . We lack an intuitive sense, an instinct, that tells us when Gaia is in danger . . . We are now so abusing the Earth that it may rise and move back to the hot state it was in fifty-five million years ago, and if it does, most of us, and our descendants will die.

These words featured in his 2006 best-seller, *The Revenge of Gaia*, which was a book designed to shake the world out of complacency:

> I think we have little option but to prepare for the worst and assume that we have already passed the threshold . . . We face unrestrained heat and its consequences will be

with us within no more than a few decades. We should now be preparing for a rise of sea levels, spells of near-intolerable heat like that in Central Europe in 2003 and storms of unprecedented severity. We should also be prepared for surprises, deadly local or regional events that are wholly unpredictable.

He predicted that oil, gas and coal emissions would cause more damage more quickly than anyone realised. Rather than the steady rise in temperature predicted by computer models, he warned there would be disturbing jags as positive feedbacks within the climate system pushed key components past the point of no return, such as the melt-off of the icecaps, the die-off of the forests or a collapse of biotic ocean pumps. 'We should expect climate changes of a kind never thought of, one-off events affecting no more than a region.'

This stark warning about the imminent collapse of the climate and society horrified millions. It was by far the quickest-selling book he wrote. And his words carried more weight than at any time in his life. Lovelock was now one of the world's most famous scientists and his theories were gaining credibility. The core Gaian concepts of planetary regulation, feedbacks, tipping points and planetary boundaries were now wrapped up in the fast-expanding field of Earth System Science.

In 2001, more than a thousand experts in this field – from the International Geosphere-Biosphere Programme, the International Human Dimensions Programme on Global Environmental Change and the World Climate Research Programme – adopted a Declaration on Global Change, which stated squarely that 'the Earth System behaves as a single, self-regulating system, comprised of physical, chemical, biological and human components'. Immediately after that announcement, Hadley Centre scientist Richard Betts excitedly wrote to inform Jim that the core Gaian idea had now been accepted. Margulis later reinforced this view: 'Earth System Science is no more than Gaia herself decked out in futuristic garb and made palatable to hard rock scientists, especially geophysicists.'

In retrospect many of the dire forecasts in *Revenge of Gaia* were remarkably prescient, but at the time the book stirred up a cacophony of opprobrium. Lovelock's message of doom went much further than almost any climate scientist was willing to go at that point. The public debate was still further behind, held up as it was by millions of dollars of climate disinformation from oil and gas companies. Jim was an important player who could help sway the discussion either way.

In Britain, Lovelock had become – in a popular phrase of that era – a 'national treasure'. He was made a Companion of Honour to Queen Elizabeth II, had his portrait hung in the National Gallery and was the subject of an entire exhibition in the Science Museum. (The curators were so taken by the 'creative chaos' of his Coombe Mill workspace that they initially planned to acquire the laboratory in its entirety, until they realised 'there were too many hazards, including radiation, mercury, asbestos, Semtex and just about everything you could possibly imagine'.)[1]

Martin Rees, who was the president of the Royal Society, and is now the Astronomer Royal, called Lovelock 'a hero to many scientists – certainly to me'.[2] The billionaire head of the Virgin Group, Richard Branson, claimed he wanted to practise 'Gaian Capitalism' and invited Lovelock on a space flight.[3] Even the queen of punk fashion, Vivienne Westwood, sought Lovelock's advice and asked for his help on a children's TV series she planned to fund that would explain climate change and outline possible solutions.[4]

Jim's following grew as he blended science, nature advocacy and wry observation in a way that few others would dare to do. 'Humans are about as qualified to be stewards of the Earth as goats are to be gardeners,' he noted in one lecture. 'We have to stop thinking of human needs and rights alone. Let us be brave and see that the real threat comes from the living earth, which we have harmed and is now at war with us,' he railed on another occasion.[5] He delighted in challenging conventional wisdom and common niceties. This was particularly true in private correspondence. In one email, he observed the respiration of people, pets and livestock makes up about a quarter of global carbon dioxide emissions, which led him to conclude: 'New green slogan. Stop breathing and improve your carbon footprint.'

He was no longer treated as a mere scientist; he was seen as a philosopher or a prophet. This was a mixed blessing. Much as he enjoyed stirring up controversy, he felt he could never give people the answers they wanted: 'They look to me as a guru and ask advice on saving the world. I always have to tell them I'm not sure I can help.'

The Revenge of Gaia was published against the backdrop of a febrile climate debate. There was something in it to shock every side. Sceptics, who were trying to underplay the threat posed by global warming, predictably dismissed the book as exaggerated and alarmist. Even many of Jim's friends, who were on the other, progressive side of the fence, felt he had gone too far. 'I fear he's overdrawing our despair budget,' lamented Chris Rapley, head of the Science Museum in London and former chief of the British Antarctic Survey.[6] In retrospect, however, the prediction of intolerable heat, severe storms and unwelcome surprises hitting within decades were to prove well founded. Less than two decades after the publication of *Revenge*, the UN Secretary General was warning of a 'Code Red' for humanity and climate scientists were horrified by off-the-chart leaps in temperatures.[7]

Among environmentalists, the most profound criticism of the book was Lovelock's claim that it was too late to prevent the 'fever of global heating'. Instead of mitigating the problems with wind farms, solar panels and 'sustainable development', he called for survivalist adaptation with atomic power and 'sustainable retreat' to hold-outs of civilisation in the Arctic. Many admirers in the environmental movement baulked at his passionate advocacy for nuclear energy and equally vehement opposition to onshore wind farms. Most were unaware of his work for the MoD and his previous association with Shell, so they could not understand why Lovelock's views were so entrenched.

Ever since the teenage hikes with his father, Lovelock had rebelled against anything that violated his sacred space, the English countryside. This included right-to-roam restrictions, fertilisers, pesticides, the European Union's Common Agricultural Policy and wind farms. In Dorset, he launched a campaign to prevent turbines being erected near Coombe Mill and his daughter Christine set up a group called Artists Against Windfarms.

Lovelock struggled to comprehend environmentalist fears about nuclear power because he had worked with radioactive elements — medical, military and industrial — almost all of his career. In the 1950s, he was already so familiar with these materials that he had been offered the job of head of the Radiation Biology Unit for the Atomic Energy Research Establishment. In the 1960s, he had invited nuclear scientists to conduct research at his family's holiday home in Adrigole and continued receiving fees for their use of his laboratory there until at least 1974. He had visited top-secret atomic weapons development sites in Aldermaston in the UK and Idaho in the United States and claimed to have seen the insides of nuclear bombs. In Japan, France and the UK, he advised companies and government ministers on nuclear energy — almost always advocating more. At his own home in Coombe Mill, he kept radioactive substances in the barn. Health and safety were anathema. If he had to choose between nuclear power and green politics, there was only going to be one winner.

The widening rift was apparent in his correspondence with Mary Midgley, who was a prominent philosopher, animal rights advocate and leading figure in the Gaia network. She had been a champion of Lovelockian thinking for many years. For her, the Gaia theory of life was an antidote to 'atomistic, individualistic, reductionist models', such as Dawkins's 'selfish gene'. Echoing

Margulis, she saw a correlation between how biologists describe the workings of life and how politicians justify their policies with appeals to human nature. 'The idea of Gaia – of life on earth as a self-sustaining natural system – is not a gratuitous, semi-mystical fantasy. It is a really useful idea, a cure for distortions that spoil our current world view.'

She and Lovelock were friends, but they fell out over Lovelock's contention that nuclear power was the only way for humanity to keep the lights on in the coming apocalypse. She expressed her dismay and urged him to recant. In reply, Jim could not control his exasperation:

10 December 2004
What a deeply disappointing letter. If you think as you wrote then the task I have to persuade the rest of the Greens is harder than I had ever imagined . . . you are among my dearest friends and I welcome your criticism but as a scientist I entreat you to see that the mountain of lies from the Green lobbies has diverted your thoughts from the truth about nuclear energy.

Midgley was equally aghast to find herself at loggerheads with one of her intellectual allies:

6 February 2005
I have been meaning to write and say how much it upsets me to find myself in opposition to you on a question that is as important to you as the energy issue when I am still, not just in entire agreement with your general world view but deeply indebted to your thinking for points that are central to my own.

The tone of their letters was never quite the same again.

This rift with Midgley was part of a broader rupture with many old allies and other green friends, such as Porritt. Even Margulis was pushed to the fringes. 'It's true that we have drifted apart,' he lamented in a 2006 letter to his long-time collaborator. 'But Sandy is not to blame, quite the contrary. It's mainly because our scientific approaches are different today.' In fact, Jim was slipping further from the frontline of science and no longer had access to the most advanced data.

This was not just a personal problem. Everything Lovelock did now had political ramifications because he had become such an influential figure. He was not well equipped to deal with this renown because his human skills lagged far behind his scientific acumen.

He had always tended to follow rather than lead on subjects of money and politics. Over the previous decades, his opinion had wandered free of any ideological tether: initially a left-wing admirer of George Orwell in the 1930s, he was also an excoriating critic of Winston Churchill during the war; an enthusiastic supporter of Clement Atlee's Labour government and its promotion of nuclear power and the National Health Service while he worked as a bureaucrat for the MRC in the late 1940s; an admirer and adviser of Conservative prime minister Margaret Thatcher while self-employed as a scientist-entrepreneur in the 1980s; a font of scorn for Labour's Tony Blair due to the war in Iraq and a mollycoddling emphasis on 'health and safety' that every pore in Lovelock's body bridled against; and an anti-European in the run-up to Brexit. He kept these views largely to himself and a small circle of friends. In public, he generally avoided taking the side of one party or another. But with the climate debate hotting up, he was never going to be allowed to stay on the fence undisturbed. He was just too juicy a political target.

Lovelock had come to the notice of one of the most powerful figures on the climate-sceptic right, the former Chancellor of the Exchequer, Nigel Lawson. This astute political operator, who had been elevated to the peerage as Lord Lawson of Blaby, was a relative newcomer to the issue, but he learned quickly. In the 1980s, Lawson had built a reputation as a high priest of neoliberal economics, famed for his Big Bang deregulation of financial markets and privatisation of state industries. He adopted a lower profile as a banking executive and diet-book author during the Blair-led Labour governments around the turn of the century, then bounced back into the ideological frontline from the mid 2000s as a result of his strong stance on the climate. Some called him Britain's 'denier-in-chief'.

The battle between trained scientists and paid sceptics was by then in full swing. The fossil-fuel industry was spending billions of dollars to delay implementation of the 1997 Kyoto Protocol,[8] under which governments promised to limit greenhouse gases. Oil and gas companies poured money into newspapers advertisements to greenwash their image and cast doubt on the science behind global heating. They hired lobbying firms, public relations agencies and 'merchants of doubt' – paid scientific stooges, many of whom had previously written sympathetic journal articles for tobacco firms and other polluting industries. Lovelock, of course, had been accused of being one of them thirty years earlier during the uproar over freons and the ozone layer. Since then, he had transformed his reputation and was now seen, rightly or wrongly, as a defender of nature. But Jim did not see sides like other people and was always open to other ideas, especially if a peer of the realm, such as Lawson, came knocking.

The politician needed a scientist. Bereft of any such training himself, he looked at the climate debate through a purely economic and political prism. For him, the campaign to reduce emissions of carbon dioxide was akin to a communist plot: 'With the collapse of Marxism, those who dislike capitalism have been obliged to find a new creed,' he wrote. 'For many of them, green is the new red.'[9] The future chairman of the Brexit Vote Leave campaign cast doubt on the science of global heating, stressed the 'positive aspects' of a warming planet, bewailed the costs of an energy transition, and criticised the work of the world's leading body of climate scientists, the United Nations IPCC. But, brilliant and eloquent as he no doubt was, Lawson was out of his depth on the more technical aspects of the debate and frequently made mistakes that undermined his credibility.

What he needed was a respected expert who sat outside of mainstream climate science, where there was 97 per cent consensus on the human causes of global warming. If that person had recently been embroiled in a high-profile spat with influential environmentalists, then that made them easier to turn. If they had once worked to publicly defend chemical companies in an environmental debate, then maybe they would already be sympathetic to the cause of industry. These considerations meant there was really only one possible candidate. Lord Lawson got to work on wooing James Ephraim Lovelock and his status-conscious wife.

The introduction was made by a mutual friend, William Waldegrave, who had also served in the Thatcher-era cabinet and was now a baron in the House of Lords. On 15 October 2007, he emailed Lovelock with a challenge: 'I'm sure it's worth keeping up on Nigel Lawson . . . Nigel responds to facts and is not beyond scientific persuasion. You can do it if anyone can.' A month later, Sandy, who was handling the couple's correspondence, replied positively that Lovelock would take up the gauntlet: 'Jim and I both enjoyed meeting and talking with Lord Lawson at both of the Climate Change seminars. He is still a sceptic of course but I do feel that if anyone could at least partly convince him otherwise, Jim could. What do you think?'

Sandy had considerable faith in her husband, but this was a mismatch. The encounter pitted one of the smartest political operators of the modern age against a semi-recluse with low self-esteem and underdeveloped social skills. From this moment on, Lovelock steadily drifted towards a sceptical viewpoint. The trend was so pronounced that it looked at one point as if Jim would revert to the position he had held on the environment while working at Shell and representing chemical companies in the ozone debate. Friends were alarmed. Could the integrity of Lovelock and Gaia survive the onslaught?

Lovelock's next book, 2009's *The Vanishing Face of Gaia*, revealed the extent of the shift. It continued where the previous one left off with a dystopian vision of the climate future, but this time the author was noticeably more critical of the United Nations IPCC (a favourite target of sceptics). Once again, he put more of the blame for the world's woes on population growth and dismissed renewable energy projects and sustainable development as mere posturing. Instead, the self-declared planetary doctor (it was no coincidence that Lovelock was a fan of Doctor Who) prescribed geo-engineering – the launch of space mirrors to reflect sunlight, the dispersal of cooling clouds of particulates into the atmosphere and the installation of ocean pipes to lift cold water to the surface (an idea he was exploring at the time with Chris Rapley and Richard Branson).[10] The overall message was that the climate could be fixed.

This was the year of the United Nations Copenhagen Climate Conference, which was the most important UN gathering since Kyoto. Lawson was primed. A few weeks before it was due to begin, he set up the Global Warming Policy Foundation, which quickly became the UK's most vociferous critic of climate action. It was later revealed that the funders included oil and gas interests.

The timing was remarkable. Supporters of the fossil-fuel industry had launched a coordinated attack on the global climate science establishment. On 17 November 2009, hacked emails from leading institutions were spread around the world on Russian servers. There was nothing incriminating in the emails, but some of the language – vaguely suggesting academics knew how to 'bury information' – caused a stir that was amplified by US senators who received donations from oil companies and a network of oil-industry-friendly media and thinktanks, including Lawson's Global Warming Policy Foundation. The ensuing ruckus, which came to be known as Climategate, was one of many reasons why Copenhagen ended in disappointment.

By this time, Lawson was in frequent correspondence with Lovelock, and his influence was growing. When Climategate hit the headlines, the doubt-mongering peer insinuated to anyone who would listen that the scientists were lying, that they were manipulating data and covering up inconvenient results that disproved the climate consensus. There was scant evidence for this defamatory claim, but Jim was quick to jump to the conclusion suggested by Lawson. In an email to the peer, he wrote: 'The revelations of data adjustment are miserable. A sin against the Holy Ghost of science.'

In public, Jim was only slightly less critical. His stance distressed former climatologist friends, particularly Peter Liss of the University of East Anglia who was close to one of those under fiercest attack, Phil Jones, the director of the Climatic Research Unit at the University of East Anglia. It was not until many years later that Lovelock admitted the scandal was trumped up, and he lamented it had cost him old friends and former colleagues: 'Climategate was all lies. It was all dirt. Peter Liss of East Anglia used to work with me often and then suddenly he wouldn't talk to me ... I used to have meals with Lord Abbotsbury, who is a senior judge, and then suddenly they were shut down due to Climategate.'

Lawson then stepped up a gear in an attempt to woo his ninety-one-year-old target to talk at his foundation. His powers of persuasion were legendary. 'Nigel was a great arm twister,' recalled Matt Ridley, a Conservative ally in the House of Lords. 'It would have been a coup for him to get Jim to speak at a Global Warming Policy Foundation event.'

In May 2010, Lawson invited Jim and Sandy to lunch at the Garrick, a private gentlemen's club in Covent Garden that was established in 1831 for 'actors and men of refinement'. On lavishly upholstered leather chairs beneath red walls filled with framed paintings, he plied Lovelock with drinks and flattery. It very nearly worked. In a drunken email on the way home, Lovelock agreed to be interviewed for the website of the Global Warming Policy Foundation, 'as long as they make it clear that I am not a member or supporter'. But something – maybe the advice of a friend, or memories of the misery he suffered at being called a 'bought man of industry' – made him change his mind. The next day, he sent a more sober email that retracted his acceptance and included an apology for sending the previous one by mistake. 'It was composed shortly after our lunch at the Garrick and before I decided that I had to stay independent,' he explained.

On at least four occasions, Lawson tried to persuade Lovelock to be publicly linked with his foundation, either through interviews or permission to reproduce his Royal Society speech, or by attending in person. Lovelock politely rebuffed these advances, saying he wanted to remain independent and a 'member of the awkward squad'. Lawson persisted: 'While it is sad that you don't feel up to giving a lecture I do of course quite understand. The alternative you suggest would nevertheless be excellent and I would like to arrange that. What do you mean by informal. Do you mean unscripted? What do you mean private, i.e. no press present? I hope it is the former but it will be as you wish.'[11]

Jim said no again, but the world's most celebrated Gaian was drifting perilously close to the right-wing, climate-sceptic camp. In one email, Lovelock expresses admiration for the 'fair and calm' website of the foundation and makes a 'widow's mite' donation to its funds.[12] In another, he encourages Lawson to persuade his friends in the Conservative government to expand methane production in the UK. He also starts to echo his powerful friend's anti-European views and belief that science was driven by green ideology.

In public, Lovelock was in the midst of a protracted climate U-turn that delighted sceptics and confused supporters. In a 2012 interview with MSNBC, Jim said he had been alarmist about climate change and accused others of exaggerating the risks.

This dismayed Betts, of the Hadley Centre, who urged Lovelock to make clear where his allegiances lay: 'You seem to have created quite a stir online with your MSNBC interview ... I'm doing my best to defend you on Twitter and if you feel the need to clarify matters I'd be very happy to put a short message from you on an appropriate blog.'[13] Lovelock wrote back emphatically: 'Just to say that the blog saying I am now a climate change denier is nonsense. I am as much on your side as ever. In my latest book I will mention that.'[14] There was no falling out, but visits to the Hadley Centre became less frequent, which meant Lovelock no longer had access to the latest data and research.

Colleagues lamented that as Jim and Sandy grew older and more isolated, they began to fashion their own version of reality. 'They would sometimes get hold of a misconception or misunderstanding and inflate it between them to an extraordinary extent – until they were totally convinced it was true and would be worked up about it. This contributed to them falling out with some long-standing friends,' one friend observed. 'In their interpersonal relationship Jim and Sandy were an extraordinary feedback system.'

Lovelock's powers were undoubtedly waning. This was also a frustrating period for his secret-service activities. This work had continued well into his nineties, but he could feel his influence was not what it had been. Several years earlier, the secret military laboratory he had helped to set up in Winfrith in Dorset in the late 1960s was shut down. He had written a passionate appeal for a reprieve:

31 March 2005
In spite of its small size and budget, the group has successfully met scientific challenges that had defeated much larger laboratories on both sides of the Atlantic ... It's easy to see why distal [Defence Science and Technology Laboratory in Porton Down] are unaware of the quality of the jewel of a group they have at

Winfrith, the security classification of the work done makes normal interaction and exchange of information between laboratories impossible.

His appeal fell on deaf ears.

Lovelock had started to grow disillusioned with political leadership during the Blair administration, particularly ahead of the second Iraq War, when he said he and his secret-intelligence colleagues knew from early on that the British government's claim that Saddam Hussein had weapons of mass destruction was 'a complete fairy story'. Even so, Jim continued to be asked for advice on issues of national security and enjoyed this work perhaps more than any other, though his pay was halved and then halved again, and some senior figures started to question why such an old man was still involved.

This was also a period of personal and money problems. Lovelock had made no financial planning for old age because he never imagined he would retire. He and Sandy were living on book royalties, reduced consultancy payments from the intelligence services, and the pension from Jim's years on the MRC.

This would probably have been enough if they remained at Coombe Mill, which was fully paid for and managed by the Gaia charity. But Sandy was not as comfortable with loneliness and isolation as her husband. Each day she would walk past the grave of Jim's former wife, Helen, and she would sometimes sit on the memorial bench for her ex, David. She felt trapped with the ghosts of the past.

She wanted to move somewhere different, somewhere by the sea. On the couple's marathon hike around the coastline fifteen years earlier, they had come across a place that she liked. Matthew Cottage was an old coastguard's cottage a stone's throw from the shingle of Chesil Beach in Dorset. Jim was initially reluctant. Even though he no longer did much practical work for the secret services, he hated the thought of not having a laboratory for the first time in his adult life. He knew that buying a new home would mean less money to live on. He would have to give up the land he had been restoring for Gaia for more than thirty years. But Sandy took precedence. In 2012, they boxed up their life and moved on.

Guardian journalist Leo Hickman was there to witness the moment, and he drew a comparison with Lovelock's intellectual retreat on the climate issue:

> James Lovelock is packing up. In one sense, the celebrated scientist and environmental author is simply moving house. Boxes and piles of unsorted papers scatter the floor of his home and 'experimental station' located in a wooded valley on the Cornwall-Devon border. But, in another sense, at the age of 92, he is finally leaving his life of scientific study and invention behind; a career that included the formulation of the Gaia theory, his highly influential hypothesis that the Earth is a self-regulating, single organism. His personal effects, notebooks and equipment are being logged and archived ahead of entering London's Science Museum's collection later this year . . . After more than three decades living amid acres of trees he planted himself by hand, he and his wife Sandy have decided to downsize and move to an old lifeguard's cottage by the beach in Dorset. 'I'm not worried about sea-level rises,' he laughed. 'At worst, I think it will be 2ft a century.' Given that Lovelock predicted in 2006 that by this century's end 'billions of us will die and the few breeding pairs of people that survive will be in the Arctic where the climate remains tolerable', this new laissez-faire attitude to our environmental fate smells and sounds like of [sic] a screeching handbrake turn.[15]

There was one side of the relocation that could not be made public. Jim asked his secret-service superiors how he should dispose of the stockpile of Semtex, TNT and other bombmaking materials he had accumulated over the decades. They apparently answered: 'Blow it up. Just light a bonfire and throw handfuls of explosives onto the fire.'

The move to Chesil Beach was a financial disaster. The proceeds of the £500,000 sale of the 40-acre property of Coombe Mill went to the Gaia charity, which used the money to buy Matthew Cottage. The remaining funds were divvied out to causes suggested by the trustees, including a Gaia PhD studentship. The couple found themselves quickly short of money until they were bailed out by the £300,000 purchase by the Science Museum of all of Lovelock's laboratory equipment and his archive of more than eighty boxes of notebooks, letters, manuscripts and other documents.[16] 'We came here absolutely broke,' Lovelock said ruefully. A few years later, Sandy wrote to wealthy friends to ask for financial support.

Given such worries, disruptions, occasional bouts of ill health and his extraordinary age, it was astonishing that Lovelock found the time and energy to continue writing books, but he said he could not afford to stop. He later acknowledged that the later work was not up to his earlier standards, but there was always an audience – even if it was ever further from the one he had attracted with the first Gaia books.

Published in 2014, *A Rough Ride to the Future* walked back on his earlier warnings of climate doom. He continued to blame fossil fuels for global heating, but shrouded climate science in uncertainty, dismissed the effectiveness of computer models and heaped scorn on IPCC projections. The writing was customarily vivid, but the conclusion – that climate fears are overblown – could have come from a Global Warming Policy Foundation press release: 'I think the final outcome for humanity may be better than we fear,' he wrote. 'I am, as I hope you will have gathered by now, an optimist. I do not envisage the death of Gaia, the Earth system, in the immediate future either through human folly or otherwise.'

Climate sceptics were thrilled that the most prominent doomist of them all had publicly recanted. Matt Ridley, the Conservative peer and adviser to the Global Warming Policy

Foundation, called the book 'sparkling', 'lucid' and 'marvellous'.[17] *The Spectator* published an article headlined 'James Lovelock On Voting Brexit, "Wicked" Renewables And Why He Changed His Mind On Climate Change'.[18] Jim's older friends in the science and environment communities were dismayed. Lovelock had been dragged into the hottest political debate on Earth, and he had been burned.

But to Sandy's delight, he was, at least, still moving in esteemed company. On 14 March 2017, Lovelock was invited to the most exclusive lunch of his life – a secret gathering in the House of Lords with Lord Lawson, Viscount (Matt) Ridley and Prince Philip, the Duke of Edinburgh. There was, apparently, no agenda beyond a general sense of frustration with the climate movement, which Lawson was more than happy to foment. Many of the details of that meeting have not been made public until now.[19]

The Queen's husband was a famous nature conservationist and a co-founder of the World Wildlife Fund (WWF), though he was known to be far less enthusiastic about the movement to address global heating through a reduced use of oil, coal and gas. A few weeks earlier, Lawson had sent him a copy of a speech that Ridley had given to the Global Warming Policy Foundation, which articulated similar complaints. Prince Philip replied with a message of support and a few grumbles of his own:

> Thank you very much for sending me a copy of Matt Ridley's lecture. I greatly enjoyed reading it because, while I have accepted the evidence that the climate is changing, I have been very doubtful whether it was entirely due to the increase of reflective layers of 'greenhouse' gases such as CO_2, in the atmosphere.[20]

Prince Philip said he had urged the WWF not to get involved in the climate debate because it would distract the organisation from the goal of conserving nature. 'Unfortunately it failed to take my advice,' he wrote. He left the organisation soon after. Interpreting this as evidence that the prince was a kindred spirit, Lawson and Ridley had invited him to this covert tryst of elite climate doubters.

The meeting had to be arranged in secret. Ridley explained why: 'First, we feared an outbreak of jealousy among colleagues. It's a club, the House of Lords. You must remember not to piss off anyone. Second, Nigel and I were Conservatives so it could have been seen as a partisan event, though we didn't see it that way.' With the permission of Black Rod, they smuggled Prince Philip up to a room that had been cleared in the Speaker's private apartment where the other three guests – including Jim, who had been chauffeured in a limousine from Dorset – were waiting.

The food had to be conveyed up in a lift, which meant it was not very warm by the time it reached the table. Fine dining, in any case, was not the point of the gathering. In fact, it was not very clear what the aim was, beyond an opportunity to let off steam. 'All of us shared doubts about the direction of environmental activism,' Ridley recalled. In a blog, he described the exchanges as 'funny, feisty, fast and furious'.

The tone was convivial, the conversation filled with anecdotes of past glories and much harrumphing at the shortcoming of the modern age. At fifty-eight years old, Ridley was the baby of the group, more than three decades younger than the average age of the others. He 'made the mistake' of complimenting his lunch partners as a 'very good advertisement for old age'. At which, he said, 'they all rounded on me and told me I did not know what I was talking about, and it was horrid being old, or words to that effect.'[21]

When the topic of climate change came up, Lawson, as ever, blithely insisted fossil fuels posed no threat whatsoever, Ridley said concerns were exaggerated and Prince Philip asserted that the science should be open to challenge. Lovelock, by his own account, told them the threat of global warming was very real and that fossil fuels needed to be phased out, but he joined in with the general tone of disapproval by warning of the threat posed by wind turbines to the British countryside.

If Lawson had hoped for something concrete from this lunch, he was disappointed. According to Lovelock and Ridley, nothing was decided. No plots agreed. As Lovelock took his limo home, he felt a little bemused. 'I left the meeting not sure what it was for. It was too much agreeing with each other disagreeing with everyone else.' His abiding memory of the event was Prince Philip being asked (by Lawson, according to Ridley) why he didn't do something about his son, Prince (later to be King) Charles, who was an avid environmentalist. 'Thirty years too late for that!' the Duke is said to have replied to guffaws of laughter.

After this, there are no more emails from Lawson trying to lure Lovelock to his side. Despite the flattery, despite the constant invitations, and despite the title of his would-be seducer, Jim had held out. He later insisted he was never in any danger of allying with the 'bunch of nutcases' at the Global Warming Policy Foundation: 'Lawson was always trying to get me to join but I wouldn't touch it with a barge pole,' he claimed.

In truth, it had been a very close-run thing. Old friends said Lovelock had considered joining the Global Warming Policy Foundation's advisory board. 'He nearly did it,' revealed one scientist who asked to remain nameless. 'I and others appealed to him not to do it. But it appealed to the maverick in him. Had he done that, I don't think he appreciated that he would become persona non grata with so many people who really admired him and who he depended on for support.'

Providence did not just smile on Lovelock at this point, she also made a joke at his expense. For in this Brexit-supporting, academic-taunting, activist-mocking phase of Jim's life, who should become the greatest global champion of his Gaia theory but the French professor of political arts Bruno Latour. Given the company Lovelock was keeping at the time, it may have seemed an unlikely match, but the two men had so much in common that it was a surprise their paths had not crossed earlier.

Latour, who was memorably described as 'France's most famous and misunderstood philosopher',[22] had spent much of his career challenging the superiority of science. Like Lovelock, he was an unorthodox, post-humanist thinker who styled himself an outsider and delighted in breaking academic silos by cross-pollinating ideas from science, the arts and religion. He had authored some of the sharpest critiques of 'modernism' – the idea that humanity had cut itself off from nature during the seventeenth-century Enlightenment and moved into a realm of culture. Also like Lovelock, he believed situations were determined by shifting networks of relationships, rather than the behaviour of discrete elements. From here, it was not a big step to embrace Margulis's insistence that organisms help to make their own environments or Lovelock's Gaian assertion in the 1990s that 'the stable state of our planet includes man as a part of, or partner in, a very democratic entity'.[23] Latour jumped on this. For him, Gaia was the politics of living things.

Latour argued that Lovelock and his ECD revealed the limits of the Earth's habitability, a discovery that was as much of a cosmological upheaval as Galileo's use of the telescope to open up ideas about the infinity of space. Gaia's cosmological move inwards, he believed, would force humanity to recognise it has to live inside the Earth in a more ecological and ethical way. 'I have done everything I can to make this accepted,' Latour said. 'Gaia is not just one more concept. It is not just about physics and energy. It is Life.'

His political approach revitalised the forty-year-old theory, shredding woolly clichés about 'Mother Earth', 'superorganism', 'harmony', 'balance' and 'unity' and restoring Gaia as an agent for change. In a series of lectures in Edinburgh in 2013 and books,[24] Latour explained how the theory had emerged from the violent technological rupture caused by industrialisation, the conquest of space, nuclear power, computerisation, cybernetics and Cold War military rivalries. Rather than a human-dominated Anthropocene, he argued this had created a 'new climate regime' in which humanity had to find a new way of reconnecting with nature to forestall calamity. It was the very opposite of Lawson's approach: while the Tory grandee hoped to use Lovelock to defend the fossil-fuelled status quo, the French philosopher believed Gaia could be used to attack a ruling elite who were using the climate crisis to wage class war and escape a planet they were leaving in ruins.

There were parallels here with the 1970s, when Gaia was being pulled one way by Rothschild, Epton and McCarthy, and another by Margulis. As then, Lovelock was delighted with the flattery he was getting from both sides. He was more than happy to describe both Latour and Lawson as friends.

Jim concentrated now on activities that were dearer to his heart: daily walks along the coast with Sandy, consulting for the intelligence services, writing a tender essay about his father, Tom, and the English countryside, and working with the author Bryan Appleyard on a new book. *Novacene* painted a techno-optimistic vision of a future dominated by a Gaia-friendly artificial intelligence. Lovelock was gratified that the book sold well, particularly in Asia, but felt his contribution was hampered by illness.

Novacene was published in July 2019, as part of the celebrations for Lovelock's 100th birthday that month. This was the occasion for a great outpouring of affection from near and afar. Newspapers and magazines celebrated his life and ideas. Exeter University ran a centenary event with three days of lectures. And, of course, there was to be a party to end all parties. Even Lovelock, who often found large gatherings wearisome and vexatious, knew this had to be done, if not for him, then for his wife.

Sandy, the consummate organiser, threw herself into the organisation of the grand event as if it were her life's great work. There was only one possible venue: Blenheim Palace, the stately home where she and Jim had locked eyes on that first night of passion and laughter. As a devoted student of military history and an Anglophile who loved England more than many English people, she was thrilled at the prospect of once again commandeering the home of Britain's wartime leader Winston Churchill. More than 100 guests would dine in the Orangery, looking out onto the 2,400-acre grounds designed by Capability Brown. Pulling together the funds would be a challenge, but she knew who she could quietly ask for a contribution. Jim needn't know. Besides, who wouldn't want to help? It would be a party for the ages.

And so it proved. On the morning of the big day, Jim had played up a little while she was helping him to get dressed. This was only to be expected. He was always a little crotchety before social gatherings. It was a sign of nerves. The tantrum had dissipated by the time they arrived at the palace, just as she knew it would. Once the guests started to arrive, the twinkle was back in his eye. She could not have felt more gratified. What a life they had enjoyed together! What friends they had made!

And here they all were to pay homage, a procession of goodwill through the decades. One by one, their cars rolled up the broad drive and the guests lined up to greet Jim, who was sitting on a chair in the orangery-like greenhouse, with a beaming smile and a welcome handshake for everyone. The attendees were a truly eclectic mix from across the world, all levels of society, all sides of the political divide. This was a testament not just to Jim's work but to his personality. As one of them, the former *Guardian* journalist Tim Radford put it, Jim 'was one of the most likeable people I've ever met, so I'm not surprised people who might consider themselves in political enmity also liked him so much'.

Sandy's eye was drawn to Lady Tickell and Sir Crispin, who had done more than anyone to lift Jim's theories to the heights where they belonged. What a great man and friend he was! Such a shame Margaret Thatcher was no longer alive. She would certainly have graced the occasion. Hadn't Jim contributed to that triumphant speech at the United Nations? Lord Lawson ought to have been here too. He was invited and she was absolutely sure he was going to attend as he had been so complimentary of Jim, calling him the most brilliant man he had ever encountered. Unfortunately, the poor fellow was recuperating from an illness in France, so he had sent his regrets and a bottle of cognac. Neither she nor Jim were drinkers any more, but it was the thought that counted.

Baron Waldegrave and Viscount Ridley had made it. She smiled at the thought of the coalmine-owning Conservative politician and that other climate sceptic at the party, James Delingpole of *The Spectator* and *Telegraph*, rubbing shoulders with the famous environmentalists on the guest list, such as the Australian conservationist Tim Flannery, the former director of Friends of the Earth, Jonathon Porritt, and Satish Kumar and Stephan Harding from Schumacher College in Dartington. Typical of Jim to bring opposites together like this. Taking sides was never his thing, but he loved to stir things up.

Stephan Harding was clearly moved by the occasion and later said:

> It was a lovely atmosphere. Very calm, friendly and elevated. All sharing spiritual champagne. This brought the best of humanity together. And all around Gaia, around Jim and Sandy. The magic that happened around them is even bigger than they are. Everyone shared it. Even the people from the Ministry of Defence. It was an image of the Earth. Under the dome, for a while, we were like a Biosphere 2 of what humanity could be like with Gaia, everyone contributing to a vision of Gaia, of being at home with the planet. We were in a greenhouse, it was a greenhousing of the Gaia human consciousness. All brought to life in the Orangery at Blenheim palace for Jim's birthday.

As guests took their seats at the brown tables filled with silver crockery and decorated with flowers, Sandy was glad Dian Hitchcock was not around; she knew her husband still harboured nostalgia about their shared history! But she wished Lynn Margulis could have been in Blenheim for Jim, as she had been for his ninetieth. Despite the many squabbles between the two great Gaia collaborators, they had been a formidable duo. Lynn had died in 2011, but her spirit was present in the form of her long-time partner, Ricardo Guerrero, a biologist at the University of Barcelona, and his friend Anna Omedes, the director of the Gaian-inspired natural science museum in Barcelona. Lord Rothschild would have been a fine guest too. Despite the chorus of detractors after his death in 1990 and the allegations that he had been a double-agent for the Soviets, Jim always defended him. One of the things Sandy admired about her husband was that his loyalties were to friends rather than tribes.

The South African concert pianist played Chopin. One mesmerised guest said that she looked like a character from a séance in the 1920s. The family were there too, of course, Christine, Jane, Andrew, John and their partners. Sandy had put them on tables at the back with Jim's Irish friends from Adrigole – the O'Sullivans and O'Boyles – as they would have much to catch up on. Jim's literary agent, Elaine Steel, was chatting to the many journalists and authors who had helped to popularise Gaia: Fred Pearce, Tim Radford, Brian Appleyard, Oliver Morton and John and Mary Gribbin.

Nobody from Shell attended, but old friends Rosemary and Bill Buffington from Hewlett-Packard were over from the United States. Tetsuro Yasuda, the coordinator of the prestigious Blue Planet Prize, which Jim had won in 1997, had come all the way from Japan to pay his respects. Other guests she knew by reputation: Stewart Brand, the editor of the *Whole Earth Catalog* who had published some of the first big Gaia articles in the 1970s, and Vivienne Westwood, the queen of punk fashion who had hoped Jim knew how to save the world. If only!

The scientists at the party knew just how hard that was. Lord Rees, the Astronomer Royal, former head of the Royal Society and author of *Our Final Century*, had almost outdone Jim as a prophet of doom at one point. She was glad he had come. Apart from being close to the royals, he had been such an important supporter of Jim's work on Daisyworld. Talking of which, it was good to see the key partner on that Gaian computer model, Professor Andrew Watson, at the party. Despite the falling-out Jim had had with him some years earlier, they had once been very close.

Other luminaries of the Gaia scene were catching up with one another over glasses of champagne: Lee Kump, who had co-authored an important paper with Jim on climate regulation; Bob Charlson, who worked with her husband on the CLAW hypothesis which was a Gaian feedback loop between plankton, sunlight and temperature; Peter Liss, the University of East Anglia professor who had been a long-term collaborator; Chris Rapley, the former head of the Science Museum who shared her husband's inventive, adventurous spirit; the influential climatologist Richard Betts of the Hadley Centre; and, of course, lovely Tim Lenton of the University of Exeter, who Jim had charged with keeping the Gaia flame burning in a new era. The heir apparent.

Tim had been working recently with Bruno Latour on something they called Gaia 2.0. Sandy knew many scientists were impressed that new life and purpose had been injected into the old theory, though she didn't think Jim would ever be able to accept that anyone but him could truly interpret Gaia. It had been such a big part of his life for so long that no matter what his mind said, deep down, his heart wouldn't let go of Ma Ga.

Still, Bruno was on fine form at the party, reading out a humorous notice of his new play, *Moving Earths*, which was a reworking of Bertholt Brecht's *Life of Galileo*, with Lovelock as the protagonist who disrupts humanity's perception of its place in the universe. As Bruno said to anyone who would listen: 'It was impossible to over-dramatise the intrusion of Gaia into contemporary thought.'

When Jim spoke, Sandy could barely contain her pride. It wasn't so much a speech as a series of conversations. He was, of course, funny, as he recounted how he had recently been bitten by an adder on one of his morning hikes. It had left a couple of fang marks on the skin of his leg, but no noticeable poison. Perhaps he was as immune to snake bites as he claimed to be to electric shocks.

Self-effacing as usual, Jim asked his guests for advice on what new projects he should take on now that he finally had to give up his work for the intelligence agencies. As he put it: 'They kept on paying me until I was 100. I was a total anomaly.' That surprised some guests who were not aware that he was the longest-serving member of the British secret services. But many others present were old comrades who had worked with Jim on counter-terrorism operations, Cold War spying and bomb detection and disposal in Northern Ireland.

He left the best part to last, a declaration of his love for Sandy and gratitude for the happiness she had given him for the previous thirty years. Her heart leapt. All those years. All that work. Her Jim had truly lived up to his promise. She meant so much to him. Tears of joy came into her eyes. There was no point trying to hold them back. That would have been as good as telling the wind not to blow.

She was already looking forward to the moment when she could have him all to herself again. Her Jim was looking a little weary now. The excitement and the years were taking their toll. When the farewells were over, she clucked over him like a mother hen. 'C'mon, kiddo,' she said with tenderness. 'Let's get you home.'

Gaia

*'I can understand you, old lady.
We're both in similar trouble.'*

The stroll along Labour-in-Vain Lane felt particularly poignant. Knowing this was probably the last time I would take this path, I drank in the sensations every step of the way and reflected on how much Jim had changed my perceptions. The warm breeze blowing in off the glimmering blue waters of the Channel contained a faint whiff of the ocean, which I knew, thanks to him, was the smell of dimethyl sulphide exhaled by seaweed and algae. Clusters of blackberry flowers had started to bloom on brambles in the hedgerows, which Jim had taught me could be roughly dated by multiplying the number of woody species by 100 years. In the steep fields that ran down towards the beach, noisily grazing lambs and ewes were bathed in mid-morning, early-summer light from our solar system's star, which, he'd informed me, had grown hotter over the past few million years, but never burned the Earth thanks to regulation of the climate by life, in all its rich complexity. In the cloudless heavens above, I knew the discharges of billions of trillions of quadrillions of bio-geo-chemical interactions all around me, within me and across the world, were constantly adjusting the colour and content and temperature of the sky. Higher still, I knew the atmosphere's cool flame was burning off the planet's excess heat and unwanted gases, though human activity, was making this ever more difficult and pushing the system towards disequilibrium. And far far up there somewhere was the red, dead planet, where Jim's ultra-sensitive gadgets were gathering Martian dust and proving that Earth alone in the solar system has the capacity for life, self-awareness and love.

Over the previous two years, I had walked this path during gentle spring rains, seen the autumn leaves paint the roadside golden brown, and trudged to the coast through mud and biting winter gales. Each time, I passed the concrete Second World War bunkers overlooking the pebble beach. Each time, I caught a glimpse of the Gaia statue nestling among the garden shrubs. Each time, I entered the cosy hobbit hole of Matthew Cottage to a cheery welcome from Jim and Sandy. Then I would snuggle into a chair, turn on my voice recorder, open my notebook and begin another session of time travelling with a man who seemed to be a Doctor Who, Ezekiel, Galileo, Gandalf, Chauncey Gardiner, Q and the Goblin in the Gasworks all rolled into one.

In that 'little stone house, deep in the countryside, away from all distracting influences', we were surrounded by mementos of a life devoted to Life: walls covered with framed 'looval' doodles, Jim's portrait of his son John, a photo with Queen Elizabeth, a panorama of Blenheim Palace and all the participants of the spiritual conference where he first met Sandy. The bookshelves contained works by Margulis, E.O. Wilson, Richard Dawkins, a biography of Margaret Thatcher and countless science fiction novels (including two he had co-written) while on the cabinet sat a surprisingly small ECD that Jim had hand-crafted on his lathe.

Most of the time Jim had been in remarkable spirits – a testament to his iron constitution and advances in modern medicine. He had been treated eight times for skin cancer, undergone open-heart surgery, lost a kidney, been afflicted with a urinary infection, suffered various bronchial ailments and spent much of his career handling toxins, radioactive substances and explosives. He relied on a hearing aid to understand what others were saying and cannabis-based pills to deal with chronic pain.

Thanks to these many interventions, he had lived twice as long as expected when he was born in 1919. Who could doubt science had furthered human progress?

Despite his many ailments, he came across as serenely happy. 'After you turn 100, people expect you to go downhill, but these have been the most content years of my life,' he insisted. He put this down to his marriage to Sandy. They had been together more than thirty years since that first night of sex and limericks in Oxford. She was his lover, his carer, his archivist and gatekeeper, managing a constant stream of media requests, product-endorsement proposals and correspondence with family and friends from a cubby hole near the back door. She was as attentive, protective and caring as any mother hen. 'Are you all right, kiddo?' 'You need to dress up warm, kid,' she would often say to the centenarian, constantly supplying us both with tea and biscuits in between internet surfing and perusal of her beloved books on military history. They appeared to be devoted to one another and did not seem to need anyone or anything else.

On several days, the three of us walked together – his lifelong favourite activity. He was in his element in nature, pointing out the different rare species that were endemic on this stretch of the Jurassic Coast and bemoaning visitors who littered the path. One time, we came across a car that had been thoughtlessly and illegally parked on a scarce species of wildflower on the edge of the beach. 'I wish I still had my stockpile of explosives,' Jim grumbled with a mischievous twinkle in his eye. 'I'd soon get rid of that.' At other times, he was frailer and needed to take my arm if there were gusts of wind. On some days, he lacked the energy to go out at all and, though he never complained of fatigue or pain, I would call a halt to our sessions after less than an hour.

There was no longer any way to ignore his own vulnerability or the planet's. 'Gaia and I are both in the last 1 per cent of our lives,' he told me. 'I have an affiliation with Gaia. I can understand you, old lady. We're both in similar trouble.' Humanity's progress had come at her expense. During Jim's lifetime he had witnessed the dramatic destabilisation of the Earth system because of industrial farming, deforestation, chemical pollution and fossil fuels, which blocked the planet's heat-waste disposal system with a blanket of human emissions. When Jim was born in 1919, the amount of carbon dioxide in the atmosphere was 303 parts per million. This rose to 320 in 1966, when he first alerted Rothschild to the risks posed by petrochemicals, and it now stood at 417, the highest level in 14 million years.

Lovelock had once again become something of a prophet of doom. He remained a believer in human ingenuity and the transformative power of technology, but he had turned against the oil industry that used to employ him. 'It's all bloody obvious. The worst thing about humanity is that those with money and power are partly aware of the dangers of fossil fuels but they don't want to risk their pensions . . . The oil and gas industry sabotaged Gaia.' Unless humanity can learn to live in partnership with the Earth, he believed the rest of creation would, as part of Gaia, unconsciously move the Earth to a new state in which humans may no longer be welcome. Covid was an example, he said, of such a negative feedback: 'Gaia will try harder next time with something even nastier.'

His ideas about a living Earth were extremely important, not because they were new in the history of human thought (which they weren't), but because they provided a scientific rationale for classical beliefs and traditional wisdom. I received ample reminders of this from Indigenous intellectuals in my Amazon rainforest home, where most of this book was written. Davi Kopenawa Yanomami, Raoni Metuktire and Patricia Gualinga were among the many Indigenous activists who explained that their communities had long considered forests, rivers and the Earth itself to be forms of life. They did not need data from atmospheric chemistry to be convinced the Earth was breathing or that nature was made up of interdependent relationships. They lived these beliefs, which had been handed down for generations. They considered themselves part of the living forest, neither separate nor superior. When I asked Lovelock whether there was a crossover between Gaia and Indigenous wisdom, he concurred with the idea. For him, the key to sustainability was a recognition that humanity was part of the planetary system—or as subsequent Gaian philosophers averred, that we live "in" not "on" the Earth.

Having read thousands of pages of his private correspondence, I knew he often tacked towards telling people what they wanted to hear. This may be human nature, but I wondered whether this tendency might be more pronounced in someone who grew up alone and then had to make up lost ground to impress his peer group. It was a different matter with data, experiments and scientific theories, when he would fight his corner bravely and brilliantly. That was the basis for his reputation as a feisty maverick. But when it came to politics, money and social interaction, Jim could sometimes come across as a somewhat gullible genius. He was vulnerable to flattery and manipulation. Even his former critic, Ford Doolittle, described Jim as 'completely without guile'.[1] Lovelock recognised this, up to a point. He told me that he wouldn't have made it so far in life if he was completely naïve. But he liked Robert May's description of him as a 'holy fool' and cherished his own capacity for 'childlike wonder'. This quality may explain his appeal to both sides of the political divide, and to both industrialists and environmentalists. It does not, however, excuse his role in some of the darker episodes of industrial history.

For better or worse, Lovelock's life illustrates the ups and downs of science in society over the past century: its capacity to improve human health and well-being, its intermeshing with the military, its promise of unbounded expansion into space, its commercial take-over by private industry to supercharge the economy, and then its revelation of its own limits and consequences, from pollution and acid rain to ozone depletion and climate disruption. This, in turn, led to environmental doubts and attempts to rethink humanity's relationship with the Earth, which threatened corporate profits and turned science into a battleground for the politics of doubt and denial. If Lovelock's life does not follow an entirely uplifting narrative arc, it is because industrial capitalism has failed to offer a happy ending. If he sometimes omitted details or misled about events, it is because the truth was distressing or shameful. If he sometimes contradicted himself, it was because he was ultimately unable to reconcile his love of industry and his concern for nature.

Lovelock was everything that was right and wrong about the twentieth century. He was a curious, funny, kind, dynamic, ambitious, jet-setting exemplar of 'progress' who was never afraid of risks, pushed himself to the limit, and felt supremely confident of his ability to invent solutions to just about every problem. Yet he was also reckless, secretive, occasionally callous, prone to low self-esteem, inclined to exaggerate his own achievements, and seduced by power, praise and knowledge. He constructed a protective myth around himself as being an independent scientist. He even came to believe it. But the contradictions forced him into evasions and deceptions.

Can we trace back through his long experience on the frontlines of science to find missteps by humanity or lessons that could help with our current predicaments? Only up to a point. Lovelock's life illustrated the pitfalls of believing too blindly in the innate virtue of industry. The marriage of science and capital had turned knowledge from something shared for the public good into something kept secret for military or commercial advantage, with often dire consequences for planetary health.

Lovelock was among the first to prove these risks, but he was reluctant to accept the Earth system was vulnerable and needed radical change in corporate behaviour and government regulation. His inertia on this existential subject suggested our species is not as nimble in its thinking as many would like to believe. Yet when he did come round and start to declare himself an environmentalist and a shop steward for other species, he made an enormous impact. That period in the late 1980s and early 1990s, when the world came together to tackle the ozone hole, organise an Earth Summit and authorise the United Nations to act on climate and biodiversity, can either be celebrated as an example of what can be done when political will is combined with scientific advocacy, or rued as the last great, missed opportunity for transformative change.

More positively, Lovelock was an important advocate for other species and the idea that an animate Earth can act as an antidote for an otherwise grimly mechanistic view of the world. He demonstrated that nature is climate and climate is nature. And he thought about the world in terms of interdependent exchanges rather than individual trajectories. This idea of life as a hub of relationships can have a practical application today in the communication of science. Lovelock – along with Margulis, Hitchcock and Latour – demonstrated the power of combining scientific facts with compelling narratives and awe-inspiring metaphors. This does not need to be how scientists talk to one another, but it can be a useful skill when addressing a wider public and trying to counter the doubts and lies spread by anti-science populists. Persuasion needs to be at an emotional as well as factual level.

Jim instinctively knew that, though he was a hesitant agent of change. He was, after all, a man of his times who had risen to prominence in a predominantly white, male, colonial, anthropocentric world. He may have been more inclined than most of his peers to listen to other voices with humility and an open mind, and he may have declared himself independent, but he was deeply embedded in – and fundamentally sympathetic to – the military-industrial complex of a patriarchal capitalist establishment.

He would never have dreamed of challenging that. He was fiercely loyal to the nation, the royal family, his comrades in the secret services, and those he considered his friends — Rothschild, Margulis, Tickell, Lawson, Latour and the chemical-industry executives. He expressed remorse that he had not done more to warn the public about the dangers of leaded petrol. I twice asked him if he also regretted not alerting the wider world to the dangers of global heating at an earlier date. The first time, he replied: 'Oh no, I can't be responsible for the planet. I just do my best.' On the second occasion, he was gently defensive: 'Look, mate, I'm only human . . . Early in the Gaia research I wouldn't have dreamed of making assumptions.'

He was almost certainly Britain's longest-serving spy. In his files was a framed hand-written letter written in green ink and signed 'C', the code name for the head of MI6: 'We are greatly in your debt for the way in which you still support our operations. The fact that this work is of seminal importance in understanding the climate means that I can thank you as well as an occupant of the planet!' Another gift, from the 'Secret Intelligence Service, Vauxhall Cross', was a drawing of a solitary man staring at a darkened street from behind the curtain of a window; wedged in the drawing's frame was another letter from 'C' dated October 2014 thanking Lovelock for fifty years of service: 'The nation is indebted to you. Yours in admiration and respect.' Colleagues told me Jim had quietly saved countless lives through his work on anti-terrorism operations.

He could be dispassionate about human suffering. Jonathon Porritt, the former director of Friends of the Earth, recalled hearing a Lovelock speech in which he expressed optimism about the future because climate catastrophe would reduce the human population by 4 billion people. 'It was said with charm, but no regret. The audience was shocked,' Porritt recalled. 'He could come across as downright insensitive.' This was the flip side of a lack of concern for his own personal safety, or, more broadly, a sense that no individual is particularly important in the grander scheme of things. This seeming indifference may also have been a coping mechanism. Almost everyone he had known had died. In the course of writing this biography, at least four of his close friends passed.

He had spoken of his own mortality in similar terms. Rather than something to be feared, it was a process, a change of state: 'To die is to be part of Gaia. All atoms mixed with the rest, except the hydrogen of course, which escapes into space.' When I recounted how my heart had stopped for more than a minute, it stirred not sympathy but curiosity: 'So what's it like on the other side?' The proximity of his own demise was another object to study. He seemed more fascinated by this than afraid. He shared calculations on the diminishing likelihood of surviving each year beyond the age of 100. There was nothing in the least bit morbid in his tone. He might just as well have been describing the margin of error in an experiment on hamster lipids. He felt our relatively short individual human lives were demonstrably and quantitatively not of significant consequence in the context of the community of life that had persisted for almost 4 billion years.

The diminishing importance of the individual was the subject of the next book he planned to write. He told me it would show how insect-like matriarchal societies were the future of evolution. 'I want to see where humans are going biologically. Spiders and ants started out as individuals in which males and females are equal. Then, when they began gathering in groups, the women took over. Most creatures that live on the ground are organised in nests dominated by women. These nests self-regulate like Gaia.' He was intrigued by both the idea and his ability to conceive it. 'That knocks a hole in the idea that you get no original thoughts after 100 years old!'

Gaia, though, was not the monopoly of one man. While Lovelock has no doubt been the most important figure in developing that conceptual framework and carrying it forward, his life demonstrates this theory is so much more than the brainchild of a single individual. Like Gaia itself, it is a group effort, a product of relationships. The credit for its Martian origins should be shared at least equally with Dian Hitchcock; the exposition of its bacterial mechanisms must be largely attributed to Lynn Margulis; the Daisyworld computer modelling that helped to convince its doubters leaned heavily on the work of Andrew Watson; and the credit for important interpretations of Gaia's meaning and usefulness needs should be shared with Tim Lenton, Bruno Latour, David Abram, Mary Midgley, Bruce Clarke, Sébastien Dutreuil and many others, including former critic Ford Doolittle, who has spent the past decade trying to 'Darwinise Gaia'. Countless contributors, including Christine and Andrew Lovelock, played less well-known roles in data collection and technological development. Still more, including Stewart Brand helped to popularise it. Others, like Rothschild, Epton, McCarthy, the secret services and chemical corporations, played a shadier but important role.

Gaia was never fixed. It is a contested, dynamic space. No one person or company dictates the shape it takes. At different times, it has been a useful excuse for oil companies, the inspiration for environmental thinking and a manifestation of planetary solidarity. If the theory teaches us anything, it is that we are all active participants in a vital, constantly evolving interdependency. Even Lovelock cannot have the final word on Gaia.

Despite a hermetic reputation, Jim's great contribution to the theory was the identification of relationships: 'My role has been to bring separated things and ideas together and make the whole more than the sum of the parts.' Curiosity drove him. Accuracy delighted him. But it was always about feeling as well as data. That is why Gaia continued to have appeal and relevance. 'Your unconscious mind can handle concepts that your conscious mind cannot handle nor speak in language. It's rebellious,' he said. Deep down he never stopped believing that the goddess was more than a metaphor. Right to the end, he was convinced the planet was truly alive.

As I approached the cottage, I felt the usual anticipation of pleasure at meeting Jim, mixed with a new feeling of unease. Sandy had warned me in advance that he was struggling to recover after a bad fall. His daughter Jane had been more candid in advising me that it might be a good moment to say goodbye. The cottage was spotlessly clean, but slightly more disordered than usual. I recalled one of Jim's earlier aphorisms: 'Clever people never have tidy places. Tidiness is the aim of the bureaucrat.' But this was different. Sandy had been too occupied caring for Jim to do other housework. Jim was frailer than I had ever seen him. He recognised me, but for the first time, his mind wandered beyond my reach.

The conversation was fragmented, concentration almost impossible. I soon gave up any thoughts of an interview and just wanted to make him feel at ease. I raised his favourite stories about walking with his father, Tom, but he no longer seemed able to look back. He appeared to be moving on, talking of things I could not see and events I could not imagine. Despite the confusion, I felt a strong sense that, even now, he was intrigued by the new phenomenon he was experiencing.

I was perturbed but did not want him to sense anything that might stress or worry him. So I stopped trying to talk about his health or his past and instead described my walk through the countryside to get there in exaggeratedly loquacious terms: the freshness of the sea breeze, the leafy green hedgerows against a crystalline blue sky, and the delicious warmth of the morning sun. It was, I said, only partly exaggerating, a perfect English summer's day.

At this he perked up noticeably and the mischievous twinkle returned to his eye as he said he would like to see that. He had not been out for two weeks and was not really supposed to be walking about in his condition, but the possibility of a rebellion had animated him. I offered him my arm and we slowly made our way out the back door, prompting an astonished grin from Sandy, who followed us out. Just a few more steps and we were in sight of Chesil Beach and the sparkling waters of the Channel. Jim smiled in the sunshine and breathed it all in, before observing: 'This is lovely, but it is not as hot as you said it was,' which made me laugh.

We talked for a few minutes more, about what, I don't remember, then I took a few photographs of Jim smiling and Sandy laughing as she covered her face because it had not yet been made up. After that, I offered to take him back inside. Sandy insisted she would do that a little later as Jim seemed to be enjoying the change of air. It was time to say goodbye. Jim left me with the exhortation: 'Live your life well.'

I cannot recall my exact reply, some mix of farewell, fond thanks and best wishes for a speedy recovery. Words seemed inadequate. We both knew we would not see one another again. I took a few steps down the path. When I looked back, Sandy was cooing around Jim like the dove he had called her in those early letters. When I looked forward, there was the sea, as seemingly serene as when I had arrived, and the sky, that great emanation of existence, deceptively unperturbed.

Notes

Nellie's diary and Lovelock's love letters were provided by Jim from his personal archive. Christine shared her diary entries and her mother's letters. My thanks to Margaret Thatcher Foundation archivist Chris Collins for information on the marginalia by Margaret Thatcher and Charles Powell on Lovelock's speeches (in the 'Sandra' chapter).

Unless otherwise stated, all documents are from the Lovelock collection at the British Science Museum, and all quotations are from interviews with the author conducted between July 2020 and June 2022. Online sources were accessed between June 2021 and December 2023.

James

1. Speaking at the Cambridge Science Festival, 2019.
2. See Edge: 'Chapter 7 "Gaia is a tough bitch"': <https://www.edge.org/conversation/lynn_margulis-chapter-7-gaia-is-a-tough-bitch>
3. Doolittle, 'Is nature really motherly?'
4. Bond, 'Exploring our love/hate relationship with Gaia'.
5. Bond, 'Exploring our love/hate relationship with Gaia'.
6. British Dyslexia Association's guide to dyscalculia: <https://www.bdadyslexia.org.uk/dyscalculia>

Nellie

1   Evans, *Eric Hobsbawm*.

Thomas

1   Wilde, *The Ballad of Reading Gaol*.
2   Lovelock, Preface to *The Natural History of Selborne*.
3   Lovelock, Preface to *The Natural History of Selborne*.

Helen

1   Letter from Roger Delahunty to Lovelock, 24 March 1986.
2   Jim's daughter Christine disputes the claim that their mother was 'anti-religion'.

Christine, Jane, Andrew, John

1   Thompson, *Harvard Hospital*.
2   'UK smallpox research could continue at Porton'. *Nature*, 277, 77 (1979).
3   Polge et al., 'The Preservation of Bull Semen'.
4   James and Lovelock, 'A preliminary investigation of the fatty acid composition'.
5   Lovelock and Smith, 'Studies on Golden Hamsters'.
6   Lovelock, 'Diathermy apparatus'.
7   Morris and Staubermann, 'Regional Styles in Pesticide Analysis'.

Dian

1   Cooper, *The Search for Life on Mars*.
2   Cooper, *The Search for Life on Mars*.
3   Letter from Hitchcock to Lovelock, dated 19 August. No year given.
4   Cooper, *The Search for Life on Mars*.
5   Letter from Fellgett to Lovelock, 1 August 1966.
6   Letter from Helen to Lovelock, shared by Christine Lovelock.

Notes

7 Lovelock and Hitchcock, 'Detecting Planetary Life from Earth'.
8 *The Hartford Courant*, 14 May 1967.
9 Cooper, *The Search for Life on Mars*.
10 Cooper, *The Search for Life on Mars*.
11 Interview with Norman Horowitz, 1984: <https://oralhistories.library.caltech.edu/22/1/OH_Horowitz_N.pdf>
12 Lovelock and Giffin, 'Planetary Atmospheres'.

Victor

1 Letter from Rothschild to a Dr Howard, 4 May 1965.
2 Invention Record, 'A Method of Tracing by Chemical Coding Procedures'. Undated.
3 Wilson, *Rothschild*.
4 Letter from Rothschild to Lovelock, 10 February 1966.
5 Lovelock, 'Midwife to the greens'.
6 Ettre and Morris, 'The Saga of the Electron-Capture Detector'.
7 Donovan, 'Roll call of Shell toxic brands deadly to insects, crop pest AND humans': <https://royaldutchshellplc.com/2010/08/20/shell-pesticides-herbicides-fungicides-and-insecticides/>
8 Morris, *From Classical to Modern Chemistry*.
9 Donovan, 'Roll call of Shell toxic brands deadly to insects, crop pest AND humans': <https://royaldutchshellplc.com/2010/08/20/shell-pesticides-herbicides-fungicides-and-insecticides/>
10 Confidential Shell Group Research Report. TLGR.0012.74. 'Concentration of Aldrin, Dieldrin and their Photoisomers in Air Samples Taken During July, 1973 in Southwest Ireland'. Lovelock is not named in this document, but it is extremely unlikely there was another Shell-linked laboratory in Bantry Bay apart from his.
11 This paper, titled 'Lead', was followed by a second report

on 16 September 1967 called 'Lead II' which noted that the measurements were taken at Shell's Thornton Research Centre.
12 Rothschild, *Meditations*.
13 Rothschild, *Meditations*.
14 The idea of self-regulation went further than anything he had discussed with Dian. In an interview with the author, she said she did not agree with the idea of self-regulation, though she did not discuss that with Jim.
15 Letter from Rothschild to Lovelock, 24 January 1967.
16 As he revealed in a letter to Stephen Schneider on 11 January 1989.
17 Rothschild, *Random Variables*.
18 Wilson, *Rothschild*.
19 Lovelock, 'Air Pollution and Climatic Change'.
20 Moore, *Margaret Thatcher*.
21 Moore, *Margaret Thatcher*.
22 Rothschild, 'The Future', address at Imperial College, 23 October 1975 (reproduced in *Memoirs of a Broomstick*).

Lynn

1 Lovelock, *Gaia: A New Look at Life on Earth*.
2 Lovelock, *Gaia: A New Look at Life on Earth*.
3 Sagan, *Lynn Margulis*.
4 Reproduced in Barlow, *From Gaia to Selfish Genes*.
5 Letter from Lovelock to Margulis, 11 September 1970.
6 Letter from Margulis to Lovelock, 24 January 1972.
7 Letter from Margulis to Lovelock, 31 March 1971.
8 Letter from Lovelock to Margulis, 17 September 1971.
9 Letter from Lovelock to Margulis, 17 April 1973.
10 Letter from Lovelock to Margulis, 15 February 1973.
11 Letter from Lovelock to Margulis, 3 January 1972.
12 William Golding journal entry, 26 October 1975.
13 William Golding journal entry, 26 October 1975.
14 Note from Lovelock to author, 6 April 2020.

Notes

15 Bruno Latour, 'The Politics of Gaia': <https://www.youtube.com/watch?v=OTeNn_mxhwk>
16 Letter from Lovelock to Margulis, 17 October 1972.
17 Letter from Lovelock to Margulis, 28 February 1973.
18 Letter from Lovelock to Margulis, 19 January 1973.
19 Lovelock and Margulis, 'Atmospheric homeostasis'.
20 Lovelock, *Homage to Gaia*.
21 Application to Shell, 1971.
22 William Golding journal entry, 24 September 1975.
23 Letter from Lodge to Maxwell, 10 December 1971.
24 Lovelock, Maggs and Wade, 'Halogenated Hydrocarbons'.
25 Andrew Watson points out that Rowlands and Molina built on work done by Stolarski and Cicerone, who published a paper in early 1974 stating that chlorine in the stratosphere would destroy ozone catalytically. However, it was purely theoretical, because they hadn't seen Jim's 1973 paper, so they didn't know there was a chlorine source in the form of the CFCs. Rowland and Molina put the Stolarski and Lovelock papers together.
26 Molina and Sherwood, 'Stratospheric sink for chlorofluoromethanes'.
27 Letter from Lovelock to Margulis, 5 June 1974.
28 Transcript of Congressional hearings.
29 Letter from Lovelock to Margulis, 7 December 1972.
30 Letter from Lovelock to Martin Ince, 19 August 1975.
31 Letter from Lovelock to Crutzen, 21 August 1975.
32 Dutreuil, *Gaïa: Terre vivante*.
33 Joseph, *Gaia: The Growth of an Idea*.
34 At a 1975 NCAR meeting to further research into Earth System Science, according to Dutreuil, *Gaïa: Terre vivante*.
35 Joseph, *Gaia: The Growth of an Idea*.
36 Letter from Lovelock to Margulis, 12 June 1975.
37 According to a letter to Stephen Schneider, 11 January 1989.
38 Letter from Lovelock to Margulis, 14 November 1972.

39 Letter from Lovelock to Margulis, 19 January 1973.
40 Email from Professor Tim Lenton to the author.
41 Letter from Margulis to Lovelock, 29 December 1973.
42 Letter from Lovelock to Margulis, 14 April 1976.
43 Lovelock and Epton, 'The Quest for Gaia'.
44 Letter from Lovelock to Margulis, 15 December 1974.
45 *New Scientist*, March 1975.
46 Letter from Lovelock to Margulis, 26 February 1975.
47 According to Epton's family in an interview with the author.
48 Letter from Brown to Epton (copied to Lovelock), 1 July 1975.
49 Letter from Fraser to Christine Lovelock, 15 November 1993.
50 Letter from Margulis to Lovelock, 12 November 1979.
51 Interview with Margulis, *New York Times Biographical Service*, 14 June 1996.
52 Margulis and Sagan. *Slanted Truths*.
53 Margulis and Sagan. *Slanted Truths*.
54 Margulis and Sagan. *Slanted Truths*.
55 Margulis and Sagan. *Slanted Truths*.

Barry

1 Lovelock, *Rough Ride to the Future*.
2 Crutzen, 'Scientists on Gaia II'.
3 Commoner, *Science and Survival*.
4 *Bomb Squad Men: The Long Walk*: <https://www.youtube.com/watch?v=y-jL3drcH_8>
5 Letter from Lovelock to Fellgett, 26 September 1969.
6 Lovelock, 'The electron-capture detector – A personal odyssey'.
7 Letter from Lovelock to Helen, 20 September 1986.
8 Snyder, *Songs for Gaia*.
9 Gould, 'Kropotkin was no crackpot'.
10 Margulis and Sagan, *Slanted Truths*.

Notes

11 Margulis and Sagan, *Slanted Truths*.
12 Cited in Kirchner, *The Gaia Hypothesis*.
13 Watson and Lovelock, 'Biological homeostasis of the global environment'.
14 Dutreuil, *Gaïa: Terre vivante*.
15 Cited in Joseph, *Gaia: The Growth of an Idea*.
16 Joseph, *Gaia: The Growth of an Idea*.
17 Bond, 'Exploring our love/hate relationship with Gaia'.

Sandra

1 'Global Forum in Oxford Faces Human Survival', *Hinduism Today*, 1 May 1988.
2 Letter from Lovelock to Sandy, 5 May 1988.
3 Letter from Lovelock to Sandy, 18 September 1988.
4 Charlson et al., 'Oceanic phytoplankton'.
5 Letter from Lovelock to David Orchard, 19 June 1988.
6 Letter from Lovelock to Garry Thomson, 18 January 1990.
7 Schumacher Lecture 1988, published in *Resurgence*. This was a big theme in public life at this time, with major campaigns by conservation groups and celebrities such as Sting. Coincidentally or not, Shell had several years earlier focused on tropical rainforests and bought some land in the Amazon.
8 Letter from Lovelock to Margulis, 4 December 1989.
9 From the John Preedy Memorial Lecture to Friends of the Earth, Cardiff, 22 September 1989.
10 From the John Preedy Memorial Lecture to Friends of the Earth, Cardiff, 22 September 1989.
11 Fax from 10 Downing Street, 4 November 1989.
12 Thatcher was dismayed by the lack of attention in the UK media, which might explain why her green enthusiasms were so short-lived: 'That speech was the fruit of much thought and a great deal of work and broke quite new political ground. But it is an

extraordinary commentary on the lack of media interest in the subject that contrary to my expectations, the television companies did not even bother to send film crews to cover the occasion. In fact I had been relying on the television lights to enable me to read my script in the gloom of the Fishmongers' Hall, where it was to be delivered; in the event a candelabra had to be passed up along the table to enable me to do so.' Thatcher, *The Autobiography*.

13 Although Thatcher did remain concerned: 'For generations, we have assumed that the efforts of mankind would leave the fundamental equilibrium of the world's systems and atmosphere stable. But it is possible that with all these enormous changes (population, agricultural, use of fossil fuels) concentrated in such a short space of time, we have unwittingly begun a massive experiment with the system of the planet itself . . . In studying the system of the earth and its atmosphere we have no laboratory in which to carry out controlled experiments. We have to rely on observations of natural systems. We need to identify particular areas of research which will help to establish cause and effect. We need to consider in more detail the likely effects of change within precise timetables. And to consider the wider implications for policy — for energy production, for fuel efficiency, for reforestation.' Thatcher, *The Autobiography*.

14 The Geological Society of London gave the prize in 2006, noting: 'Even in the illustrious history of the Society's principal medal, first awarded to William Smith in 1831, it is rare to be able to say that the recipient has opened up an entire new field of study in Earth science. But such is the case with this year's winner, James Lovelock . . . The view of the planet and the life on it as a single complex system, analogous in some ways to a homeostatically self-regulating organism, is what gave

birth to the field we now know as "Earth System Sciences", also the most recently formed of this society's specialist groups.'

15 Lovelock, 'From God to Gaia', the *Guardian*, 4 August 1999.

Nigel and Bruno

1   Ball, 'Last of the independents?', Prospect blog: <https://philipball.blogspot.com/2014/04/last-of-independents.html>
2   Foreword by Martin Rees in Lovelock, *The Vanishing Face of Gaia*.
3   Email from Branson to Lovelock, 2 November 2006.
4   Email from Westwood to Lovelock, 12 December 2009.
5   'Nuclear energy for the 21st Century', speech to the International Conference in Paris, 21–22 March 2005.
6   Goodell in 'James Lovelock, the Prophet', *Rolling Stone*, 1 November 2007.
7   United Nations press release, 'Secretary-General Calls Latest IPCC Climate Report "Code Red for Humanity", Stressing "Irrefutable" Evidence of Human Influence', 9 August 2021.
8   At least $3.6bn over three decades, by one estimate: <https://www.theguardian.com/business/2020/jan/08/oil-companies-climate-crisis-pr-spending>
9   Lawson, *An Appeal to Reason*.
10 Lovelock and Rapley, 'Ocean pipes'.
11 Email from Lawson to Lovelock, 5 May 2012.
12 Email from Lovelock to Lawson, 21 November 2016.
13 Email from Betts to Lovelock, 27 April 2012.
14 Email from Lovelock to Betts, 28 April 2012.
15 Hickman, 'James Lovelock: The UK should be going mad for fracking', the *Guardian*, 15 June 2012.
16 With support from numerous individuals, including Vivienne Westwood and Knut Kloster, according to the *Science Museum Annual Review 2012–13*.

17 Ridley, 'A Rough Ride to the Future', 8 April 2014
18 The *Spectator*, 9 September 2017.
19 Matt Ridley mentioned the gathering in the House of Lords after Prince Philip's death and later wrote a short article in the *Spectator*, but he barely touched on the secrecy involved until his interview for this book.
20 Email shared by Matt Ridley.
21 Ridley, 'My Unexpected Lunch with Nigel Lawson – and Prince Philip' < https://www.mattridley.co.uk/blog/my-unexpected-lunch-with-nigel-lawson-and-prince-philip/>.
22 Kofman, 'Bruno Latour, the Post-Truth Philosopher, Mounts a Defense of Science', the *New York Times*, 25 October 2018.
23 Lovelock, *Gaia: A New Look at Life on Earth*.
24 For example, *Facing Gaia*, *Down to Earth* and *After Lockdown*.

Gaia

1 Doolittle, 'Making Evolutionary Sense of Gaia'.

Bibliography

Books

Barlow, Connie. *From Gaia to Selfish Genes: Selected Writings in the Life Sciences*. MIT Press, 1991.

Bell, Alice. *Our Biggest Experiment: A History of the Climate Crisis*. Counterpoint LLC, 2021.

Carson, Rachel. *Silent Spring*. Houghton Mifflin, 1962.

Clarke, Bruce and Dutreuil, Sébastien. *Writing Gaia: The Scientific Correspondence of James Lovelock and Lynn Margulis*. Cambridge University Press, 2022.

Commoner, Barry. *The Closing Circle: Nature, Man and Technology*. Knopf, 1971.

Commoner, Barry. *Making Peace with the Planet*. Pantheon, 1990.

Cooper, Henry S.F. *The Search for Life on Mars*. Henry Holt, 1976.

Davies, Nicola. *Gaia Warriors*. Walker Books, 2009.

Dawkins, Richard. *The Selfish Gene*. Oxford University Press, 1976.

Dutreuil, Sébastien. *Gaïa: Terre vivante*. Empêcheurs de penser rond, 2024.

Egan, Michael. *Barry Commoner and the Science of Survival: The Remaking of American Environmentalism*. MIT Press, 2007.

Evans, Richard. *Eric Hobsbawm: A Life in History*. Oxford University Press, 2019.

Golding, Judy. *The Children of Lovers: A Memoir of William Golding by his Daughter*. Cerise Press, 2011.
Gribbin, John and Mary. *He Knew He Was Right: The Irrepressible Life of James Lovelock*. Allen Lane/Princeton University Press, 2009.
Hawking, Stephen. *My Brief History*. Bantam, 2013.
Isaacson, Walter. *Einstein: His Life and Universe*. Simon & Schuster, 2020.
Joseph, Lawrence E. *Gaia: The Growth of an Idea*. Arkana Books, 1990.
Lambright, W. Henry. *Why Mars: NASA and the Politics of Space Exploration*. John Hopkins University Press, 2014.
Latour, Bruno. *Facing Gaia: Eight Lectures on the New Climatic Regime*. Harvard University Press, 2017.
Latour, Bruno. *Down to Earth: Politics in the New Climatic Regime*. Polity Press, 2018.
Latour, Bruno. *Critical Zones: The Science and Politics of Landing on Earth*. MIT Press, 2020.
Latour, Bruno. *After Lockdown: A Metamorphosis*. Polity Press, 2021.
Latour, Bruno. *How to Inhabit the Earth: Interviews with Nicolas Truong*. Polity Press, 2023.
Lawson, Nigel. *An Appeal to Reason: A Cool Look at Global Warming*. Overlook Duckworth, 2008.
Lawson, Nigel. *Memoirs of a Tory Radical*. Biteback, 2011.
Lovelock, James. *Gaia: A New Look at Life on Earth*. Oxford University Press, 1979.
Lovelock, James and Allaby, M. *The Greening of Mars*. Andre Deutsch, 1984.
Lovelock, James. *The Ages of Gaia: A Biography of Our Living Earth*. Oxford University Press, 1988.
Lovelock, James. *Gaia: The Practical Science of Planetary Medicine*. Gaia Books, 1991.
Lovelock, James. *Homage to Gaia: The Life of an Independent Scientist*. Oxford University Press, 2000.

Bibliography

Lovelock, James. *The Revenge of Gaia: Why the Earth is Fighting Back and How We Can Still Save Humanity.* Allen Lane, 2007.

Lovelock, James. *The Vanishing Face of Gaia: A Final Warning.* Penguin, 2010.

Lovelock, James. Preface to *The Natural History of Selborne*, by Gilbert White. Little Toller Books, 2014.

Lovelock, James. *A Rough Ride to the Future.* Cambridge University Press, 2014.

Lovelock, James et al. *The Earth and I.* Taschen, 2016.

Lovelock, James and Appleyard, Brian. *Novacene: The Coming Age of Hyperintelligence.* Penguin, 2020.

Lovelock, James. *We Belong to Gaia.* Penguin, 2021.

Margulis, Lynn and Sagan, Dorion. *Slanted Truths.* Springer New York, 1997.

Margulis, Lynn and Lovelock, James. Introduction to *Scientists Debate Gaia: The Next Century*, edited by Stephen H. Schneider, James R. Miller, Eileen Crist and Penelope J. Boston. MIT Press, 2004.

Moore, Charles. *Margaret Thatcher, the Authorised Biography. Volume 3: Herself Alone.* Allen Lane, 2020.

Morris, Peter. "'Parts per trillion is a fairy tale': the development of the electron capture detector and its impact in the monitoring of DDT' in *From Classical to Modern Chemistry.* Royal Society of Chemistry, 2002, pp. 259–84.

Rothschild, Victor. *Meditations of a Broomstick.* Collins, 1977.

Rothschild, Victor. *Random Variables.* Collins, 1984.

Ruse, Michael. *The Gaia Hypothesis: Science on a Pagan Planet.* University of Chicago Press, 2013.

Sagan, Dorion. *Lynn Margulis: The Life and Legacy of a Scientific Rebel.* Chelsea Green, 2013.

Snyder, Gary. *Songs for Gaia.* Copper Canyon Press, 1979.

Thatcher, Margaret. *Margaret Thatcher: The Autobiography.* Harper Press, 1995.

Thompson, Keith R. *Harvard Hospital and Its Volunteers: Story of the Common Cold Research Unit.* Danny Howell Books, 1990.

Thomson, Jennifer. *The Wild and the Toxic: American Environmentalism and the Politics of Health*. The University of North Carolina Press, 2019.

Tickell, Crispin. *Climatic Change and World Affairs*. Center for International Affairs, Harvard University, 1977.

Tickell, Crispin. 'Gaia and the Human Species', chapter in *Scientists Debate Gaia: The Next Century*, edited by Stephen H. Schneider, James R. Miller, Eileen Crist and Penelope J. Boston. MIT Press, 2004.

Wilson, Derek. *Rothschild: A Story of Wealth and Power*. Andre Deutsch, 1988.

Wulf, Andrea. *The Invention of Nature: The Adventures of Alexander von Humboldt*. John Murray, 2015.

Academic papers by Lovelock and collaborators

In chronological order:

Bourdillon, R.B., Lidwell, O.M. and Lovelock, J.E. 'Sneezing and disinfection by hypochlorites'. *British Medical Journal*, 42 (1942).

Lovelock, J.E., Lidwell, O.M. and Raymond, W.F. 'Aerial disinfection'. *Nature*, 153, 20 (1944).

Lovelock, J.E. 'Wax Pencil for Writing on Cold Wet Glassware'. *Nature*, 155 (1945).

Lovelock, J.E. 'The properties and use of aliphatic and hydroxy carboxylic acids in aerial disinfection'. London School of Hygiene and Tropical Medicine PhD thesis (1947).

Dumbell, K.R., Lovelock, J.E. and Lowbury, E.J. 'Handkerchiefs in the transfer of respiratory infection, l'. *Lancet*, 255, 183 (1948).

Lovelock, J.E and Wasilewska J. 'An Ionisation Anemometer'. *Scientific Instruments*, 26 (1949).

Andrewes, C.H., Lovelock, J.E. and Sommerville, T. 'An experiment on the transmission of colds'. *Lancet*, 260, 25 (1951).

Lovelock, J.E., Porterfield, J.S., Roden, A.T., Sommerville, T. and Andrewes, C.H. 'Further studies on the natural transmission of the common cold'. *Lancet*, 657 (1952).

Bibliography

Polge, C. and Lovelock, J.E. 'The Preservation of Bull Semen at -79C'. *Veterinary Record*, 64 (1952).

Lovelock, J.E. and Parkes, A.S. 'Resuscitation of Hamsters after Partial Crystallization at Body Temperatures Below 0C'. *Nature*, 173 (1954).

Smith, A.U., Lovelock, J.E. and Parkes, A.S. 'Resuscitation of hamsters after supercooling or partial crystallization at body temperatures below 0C'. *Nature*, 173, 1136 (1954).

James, A.T. and Lovelock, J.E. 'A preliminary investigation of the fatty acid composition of blood lipids from rabbit, ox, rat and normal and atherosclerotic humans'. IIIrd International Conference on Biochemical Problems of Lipids (26–28 July 1956).

Lovelock, J.E. and Smith, A.U. 'Studies on Golden Hamsters during Cooling to and Rewarming from Body Temperatures below 0 degrees C. III. Biophysical Aspects and General Discussion'. *Proceedings of the Royal Society of London, Series B, Biological Sciences*, 145, 920 (1956).

Lovelock, J.E. 'Diathermy apparatus for the rapid rewarming of whole animals from 0C and below. *Proceedings of the Royal Society of London, Series B*, 147, 545 (1957).

James, A.T., Lovelock, J.E., Webb, J. and Trotter, W.R. 'The fatty acids of the blood in coronary-artery disease'. *Lancet*, 705 (1957).

Lovelock, J.E. 'A Sensitive Detector for Gas Chromatography'. *Journal of Chromatography*, 1 (1958).

Lovelock, J.E. 'Argon detectors'. *Gas Chromatography*, (ed.) Scott, R.P.W. (1960), pp. 16–29.

Lovelock, J.E. 'A physical basis for life detection experiments', *Nature*, 207, 4997 (7 August 1965), pp. 568–70.

Hitchcock, Dian R. and Lovelock, J.E. 'Life detection by atmospheric analysis'. *Icarus: International Journal of the Solar System*, 7, 2 (September 1967).

Lovelock, J.E. and Hitchcock, D.E. 'Detecting Planetary Life from Earth'. *Science*, 3, 56 (1967).

Lovelock, J.E. and Giffin, C.E. 'Planetary Atmospheres: compositional and other changes associated with the presence of life'. *Advances in the Astronautical Sciences*, 25 (1969).

Lovelock, J.E. 'Air Pollution and Climatic Change'. *Atmospheric Environment*, Volume 5 (1971).

Lovelock, J.E. 'Atmospheric Fluorine Compounds as Indicators of Air Movements'. *Nature*, 230, 5293, (1971).

Lovelock, J.E. 'Atmospheric Turbidity and CC13F Concentrations in Rural Southern England and Southern Ireland'. *Atmospheric Environment*, 6 (1972).

Lovelock, J.E., Maggs, R.J. and Rasmussen, R.A. 'Atmospheric Dimethyl Sulphide and the Natural Sulphur Cycle'. *Nature*, 237 (1972), pp. 452–3.

Lovelock, J.E., Maggs, R.J and Wade, R.J. 'Halogenated Hydrocarbons in and over the Atlantic'. *Nature*, 241 (5386): 194 (1973).

Lovelock, J.E. 'Atmospheric Halocarbons and Stratospheric Ozone'. *Nature*, 252 (1974), pp. 292–4.

Lovelock, J.E. and Penkett, S.A. 'PAN over the Atlantic and the Smell of Clean Ocean'. *Nature*, 249 (1974).

Lovelock, J.E. and Margulis, Lynn. 'Atmospheric homeostasis by and for the biosphere: the Gaia hypothesis'. *Tellus*, XXVI (1974).

Lovelock, J.E. 'Natural Halocarbons in the Air and in the Sea'. *Nature*, 256, 5514 (1975).

Lovelock, J.E. 'PAN in the Natural Environment; Its Possible Significance in the Epidemiology of Skin Cancer'. *Ambo*, 6, 2/3 (1977).

Lovelock, J.E. 'The electron-capture detector – A personal odyssey'. *Journal of Chromatography*, 20 (1981).

Lovelock, J.E. and Watson, A.J. 'The Regulation of Carbon Dioxide and Climate: Gaia or Geochemistry'. *Planet Space Science*, 30, 8 (1982).

Lovelock, J.E. 'A numerical model for biodiversity'. *Philosophical Transactions of the Royal Society B*, 338 (1992), pp. 383–91.

Midgley, Mary. 'Individualism and the concept of Gaia'. *Review of International Studies* (2000).
Watson, Andrew J. and Lovelock, J.E. 'Biological homeostasis of the global environment: the parable of Daisyworld'. *Tellus*, 35B (1983), pp. 284–9.
Lovelock, J.E. 'Gaia as seen through the atmosphere'. *Atmospheric Environment*, 6 (1972), pp. 579–80.
Lovelock, J.E. 'Geophysiology: a new look at earth science'. *Bulletin Of The American Meteorological Society*, 67, 4 (April 1986).
Charlson, Robert, Lovelock, J.E., Andreae, Meinrat O. and Stephen, G. Warren. 'Oceanic phytoplankton, atmospheric sulphur, cloud albedo and climate (the CLAW hypothesis)'. *Nature*, 326, 6114 (16 April 1987).
Lovelock, J.E. 'Geophysiology, the science of Gaia'. *Reviews of Geophysics*, 17 (11 May 1989).

Other Lovelock publications (in newspapers, magazines, etc.)

In chronological order:
Lovelock, J.E. Letter regarding role of Dian Hitchcock in the theory of life detection. *Scientific American*, 221, 1 (July 1969), pp. 8–11.
Lovelock, J.E. and Epton, S. 'The Quest for Gaia'. *New Scientist* (6 February 1975).
Lovelock, J.E. 'The Independent Practice of Science'. *New Scientist* (6 September 1979).
Lovelock, J.E. 'Are We Destabilising World Climate: The Lessons of Geophysiology'. *The Ecologist*, 15 (1985).
Lovelock, J.E. 'I Speak for the Earth'. *Resurgence*, 129 (1988).
Lovelock, J.E. 'The Greening of Science'. *Resurgence*, 138 (1989).
Lovelock, J.E. 'A Danger to Science?' (review of *The Rebirth of Nature*, by Rupert Sheldrake). *Nature*, 348 (1990).
Lovelock, J.E. 'Travels with and Electron Capture Detector'. Blue Planet Prize speech (1997).

Lovelock, J.E. 'Midwife to the greens: the electron capture detector'. *Microbiologia*, March, 13 (1) (1997), pp. 11–22.

Lovelock, J.E. 'A Way of Life for Agnostics?' *Sceptical Inquirer*, 25 (2001).

Lovelock, J.E. 'What is Gaia?' *Resurgence*, 211 (2002).

Lovelock, J.E. 'Biographic Memoir of Archer, John Porter Martin CBE'. *Royal Society*, 50 (2004).

Lovelock, J.E. and Rapley C. 'Ocean pipes could help the Earth to cure itself'. Letter to *Nature* (26 September 2007).

Related papers by other authors

In alphabetical order:

Aronowsky, Leah. 'Gas Guzzling Gaia, or: A Prehistory of Climate Change Denialism'. *Critical Inquiry*, 47, 2 (2021).

Battistoni, Alyssa. 'Latour's Metamorphosis'. *New Left Review* (20 January 2023).

Bellinger, David C. and Bellinger, Andrew M. 'Childhood lead poisoning: the torturous path from science to policy'. *Journal of Clinical Investigation*, 116, 4 (3 April 2006), pp. 853–7.

Bolin, Bert and Arrheniust, Erik. 'Nitrogen: A Special Issue' (Stockholm Nobel Symposium). *Ambio*, 6, 2/3 (1977).

Bond, Michael. 'Exploring our love/hate relationship with Gaia'. *New Scientist* (21 August 2013).

Commoner, Barry. 'Cost-Risk-Benefit Analysis of Nitrogen Fertilization: A Case History'. *Ambio*, 6, 2/3 (1977).

Crutzen, Paul. 'Scientists on Gaia II'. *Proceedings of the 2nd Chapman Conference on the Gaia Hypothesis* (2002), p. 102.

Doolittle, Ford. 'Is nature really motherly?' *CoEvolution Quarterly*, Spring (1981), pp. 58–63.

Doolittle, Ford. 'Making Evolutionary Sense of Gaia'. *Trends in Ecology and Evolution* (2019).

Dutreuil, Sébastien. 'Lovelock, Gaïa et la pollution: un scientifique entrepreneur à l'origine d'une nouvelle science et d'une philosophie politique de la nature'. *Zilsel*, 2 (2): 19 (August 2017).

Bibliography

Egan, Michael. 'Subaltern Environmentalism in the United States: A Historiographic Review'. *Environment and History*, 8 (2002), pp. 21–41.

Ettre, Leslie and Morris, Peter. 'The Saga of the Electron-Capture Detector'. *LCGC North America*, 25, 2 (2007).

Fleischaker, G.R. and Margulis, L. 'Autopoiesis and the origin of bacteria', *Advances in Space Research* (1986).

Gray, John. 'James Lovelock: A man for all seasons'. *New Statesman* (27 March 2013).

Hitchcock, D.R. and Thomas, G.B. 'Statistical Decision Problems in Large Scale Biological Experiments'. Report submitted to Office of Space Sciences, National Aeronautics 6 Space Administration (31 March 1965): <https://ntrs.nasa.gov/api/citations/19660020153/downloads/19660020153.pdf>

Hitchcock, D.R. and Thomas, G.B. 'Continuation of Studies in Statistical Decision Theory in Large Scale Biological Experiments'. Report submitted to Office of Space Sciences, National Aeronautics 6 Space Administration (31 July 1966): <https://core.ac.uk/download/pdf/85250393.pdf>

Hitchcock, D.R. 'Sulfuric acid aerosols and HCl release in coastal atmospheres: Evidence of rapid formation of sulfuric acid particulates'. *Atmospheric Environment*, 14, 2 (1967).

Hitchcock, D.R. 'Potential for atmospheric sulfur from microbiological sulfate reduction'. *Atmospheric Environment*, 16, 1 (1967).

Hitchcock, D.R. 'Evidence of biogenic sulfur oxides in a salt marsh atmosphere'. *Atmospheric Environment*, 18, 1 (1967).

Hitchcock, D.R. 'Gaia Data' (letter to editor). *New Scientist* (March 1975).

Hitchcock, D.R. 'Dimethyl sulfide emissions to the global atmosphere'. *Chemosphere* (1975).

Hitchcock, D.R. 'Atmospheric Sulfates from Biological Sources'. *Journal of the Air Pollution Control Association*, 26: 3 (1976).

Kirchner, James W. 'The Gaia Hypothesis: Fact, Theory and Wishful Thinking'. *Climatic Change* (2002).

Latour, Bruno. 'Bruno Latour Tracks Down Gaia'. *Los Angeles Review of Books* (3 July 2018).

Latour, Bruno and Lenton, Timothy M. 'Extending the Domain of Freedom, or Why Gaia Is So Hard to Understand'. *Critical Inquiry* (2019).

Lenton, Timothy M., Dutreuil, Sébastien and Latour, Bruno. 'Life on Earth is Hard to Spot'. *The Anthropocene Review* (2020).

Molina, Mario and Rowland, Sherwood. 'Stratospheric sink for chlorofluoromethanes: chlorine atom-catalysed destruction of ozone'. *Nature*, 249 (1974), pp. 810–2.

Morris, Peter and Staubermann, Klaus. 'Regional Styles in Pesticide Analysis: Coulson, Lovelock, and the Detection of Organochlorine Insecticides'. In *Illuminating Instruments*, Smithsonian Institution Scholarly Press (2009).

Needleman H.L. 'The removal of lead from gasoline: historical and personal reflections'. *Environmental Research*, September, 84 (1) (2000), pp. 20–35.

Postgate, J. 'Gaia Gets Too Big for Her Boots'. *New Scientist*, 118 (1988), p. 60.

Ridley, Matt. 'My Unexpected Lunch with Nigel Lawson – and Prince Philip'. Blog (5 April 2023): <https://www.mattridley.co.uk/blog/my-unexpected-lunch-with-nigel-lawson-and-prince-philip/>

Rosner, D. and Markowitz, G. 'A "gift of God"?: The public health controversy over leaded gasoline during the 1920s'. *American Journal of Public Health*, April; 75 (4) (1985), pp. 344–52.

Audio and video recordings

In alphabetical order:

Bruno Latour. Inside the 'Planetary Boundaries': Gaia's Estate: <https://www.youtube.com/watch?v=5xojsnUtXHQ>

Bruno Latour on the Politics of Gaia: <https://www.youtube.com/watch?v=OTeNn_mxhwk>

Bibliography

Bruno Latour: Why Gaia is not the Globe: <https://www.youtube.com/watch?v=7AGg-oHzPsM>

Cambridge Science Festival event, 2019 – James Lovelock 100th birthday discussion with Chris Rapley, Helen Czerski and Tim Lenton: <https://climateseries.com/home-blog/41-jim-lovelock-rapley-ccls2019>

Look at Life – Rising to high office, Shell Centre documentary, 1963: <https://www.youtube.com/watch?v=8zUQD1p9bXY>

National Life Stories: An Oral History of British Science. James Lovelock Interviewed by Paul Merchant, 2010: <http://www.psichenatura.it/fileadmin/img/James_Lovelock_Interviewed_by_Paul_Merchant_2010.pdf>

Robert V. Meghreblian interview. Jet Propulsion Laboratory Archives, Pasadena, California, 1971: <https://search.worldcat.org/title/733102261>

Scott J. Horowitz Oral History Interviews, NASA Johnson Space Center Oral History Project, 2007: <https://historycollection.jsc.nasa.gov/JSCHistoryPortal/history/oral_histories/HorowitzSJ/horowitzsj.htm>

Tim Lenton interviews James Lovelock for Lovelock Centenary Conference in Exeter, 2019: <https://www.youtube.com/watch?v=MGziItCwDJA>

Acknowledgements

The completion of a book is a joyful moment, but in this case, also tinged with sadness because part of me is reluctant to let go. Researching Jim's life allowed me to travel in time and intellectual space at a period when my physical movements were restricted by poor health. But this book no more belongs solely to me than Gaia belongs solely to Lovelock. I must give thanks for the extraordinary relationships that made this biography possible.

First to the life-saving National Health Service staff at St George's Hospital, Tooting, without whom I would never have had the chance to even start this journey. I am convinced we need a planetary version of the NHS – and not just for its human inhabitants.

Next to Jim and Sandy for letting me into their lives, sharing stories, jokes, tea and biscuits, walks along the coast and intimate letters and diaries. This was not just about the work. We were friends. Jim could not talk to me about certain subjects and sometimes his inventiveness extended to his version of events, but he never tried to tell me what to write and I believe he opened up as much as he could. I will cherish those cosy morning talks in Matthew Cottage for the rest of my life.

Christine, Jane, Andrew and John Lovelock were extremely generous with their time and memories. They highlighted their father's strengths, pointed out inconsistencies and filled in missing information, particularly about the loving and important relationship that Jim had with their mother, Helen. I am particularly grateful to Jane for putting me up on two occasions and to Christine for sharing extracts from her diaries and other documents and photographs from the family archive.

Bruno Latour nudged me to start this exploration with an online interview during lockdown when he argued the pandemic was an opportunity to viralise a new Gaia-centred political ecology.

Invaluable insights were also gained from interviews and email exchanges with Anders Wijkman, Andrew Watson, Charles Powell, Chris Rapley, David Abram, Ford Doolittle, Fred Pearce, Jenny Powys-Lybbe, Jonathan Porritt, Martin Rees, Matt Ridley, Michael Egan, Nicky Campbell (daughter of Sidney Epton), Penelope and Crispin Tickell, Peter Morris, Peter Simmonds, Richard Betts, Satish Kumar, Sébastien Dutreuil, Stephan Harding, Thomas Rosswall, Tim Lenton, Tim Radford and Jim's former colleagues in the British Secret Services, who spoke on condition of anonymity. Special thanks to Dian Hitchcock for confiding in me. I hope this book can go some way towards securing the recognition she deserves.

After so many decades even the sharpest memories and best intentions can lead a biographer astray, so I must also express appreciation to the institutional keepers of records: Nicola Presley and Judy Golding of the William Golding archives; Chris Collins at the Margaret Thatcher Foundation Archive; the helpful staff at the British Science Museum Archives, Jet Propulsion Laboratory Archives; the NASA archives; the London Metropolitan Archives; the Royal Swedish Academy of Sciences; and the British Library. Immense gratitude must also go to Julia Ranney in Washington for sifting through the Library of Congress archive on Barry Commoner and putting in a freedom of information request about him to the FBI.

This book could not have been published without the advice, support and encouragement of Bryony Worthington, Sophie Lambert, Max Pugh, Anna Southgate, Ewen MacAskill,

Acknowledgements

Sharmilla Beezmohun, Rebecca Carter and Mark and Pippa Cartmell. Enormous thanks, too, to the brilliant editors and agents at Canongate, particularly Jamie Byng, Simon Thorogood, Craig Hillsley and Vicki Rutherford. I'd also like to thank Jen Gauthier, James Penco and Jennifer Croll at Greystone Books for believing in this book for a North American audience and for their care and support around environmental and Indigenous issues.

It would also have many more mistakes and omissions were it not for the assiduous fact-checking of Andrew Watson, Christine Lovelock, Sébastien Dutreuil, Tim Lenton and Tim Radford. Loving appreciation is also due to Aimee and Emma Watts, Simon Lafrenais and José Brum for casting a caring eye on the early drafts and helping set me off in the right direction.

Most of all, I must thank Eliane Brum, who nursed me to health, married me, spent a month in a Dorset caravan during the research, put up with manic late-night writing bouts while we were building a home in the Amazon rainforest, and provided love, patience and wise counsel throughout the four-year gestation of this book. I wouldn't be here without you.

Index

Abram, David 209, 211, 269
Admiralty Materials
 Laboratory 116
Adrigole (Ireland) 130,
 149–50, 160–3, 200, 212
aerosols 166, 168, 171–4
Afghanistan 195
airborne disease 54, 63, 64,
 69
aldrin 121, 122, 123
Allaby, Michael 204
Ambrose Barlow Catholic
 Society (Manchester) 48–9
Amsterdam Prize for the
 Environment 232
Andrewes, Christopher 2, 63,
 66
anthrax 68
anti-Semitism 41
apartheid 217
Appleyard, Bryan 258
 Novacene 255
applied physiology 53

argon 75
Arrhenius, Svante 128
Asimov, Isaac 204
Atlee, Clement 241
atmospheric monitoring *see*
 ECD (electron capture
 detector)
atomic bomb 60, 69, 239
Atomic Energy Research
 Establishment 67, 75, 161

bacteria 153, 154
Barber, Ed 222
BBC 74, 145, 182, 209
Beara peninsula (Ireland)
 160–1
Beckton gas works 42–3
Bell Labs 81
Betts, Richard 236, 247, 259
Bikini Atoll 60
biological weapons 64, 69
Birkbeck College (London)
 52

Blackett, Patrick 52
Blair, Tony 241, 242, 248
Blenheim Palace (Oxon) 255–60
Blue Planet Prize 233
Blunt, Anthony 145
Boer, Hendrik 75
Bolin, Bert 188, 189, 192
bombs 8, 25, 31–3, 198–9, 249–50
 and atomic 60, 69, 239
 and Northern Ireland 194–5, 200
Bourdillon, Robert 54
Bowerchalke (Wilts) 63, 79, 157, 197–8
Brand, Stewart 178, 258, 269
Branson, Richard 237, 244
Braun, Wernher von 60, 81
Brazzos 195, 197
Brecht, Bertholt: *Life of Galileo* 259
Brimble, Jack 118
British Antarctic Survey 167
British Army 194–5
British Medical Journal 67–8
Brixton (London) 22–4, 25
Brown, John 182
Brum, Eliane 7
Bryce-Smith, Derek 125–7
Buffington, Rosemary and Bill 258
Burgess, Guy 145
burns 59, 60

California (USA) 81
carbon dioxide 109, 128, 131, 140, 234, 264
 and Lawson 242
 and Mars 90
Carl XVI Gustaf of Sweden, King 188
Carson, Rachel: *Silent Spring* 121–2
Cavendish, Henry 175
CCRU *see* Common Cold Research Unit (CCRU)
CFCs (chlorofluorocarbons) 5, 137, 142, 176
 and Ireland 162–3
 and Montreal Protocol 217
 and Rowland 170–1
 and southern hemisphere 166–70
 and stratosphere 171–3
Challenger, Frederick 163
Charles III of Great Britain, King 253
Charlson, Bob 259
chemical weapons 64
Chesil Beach (Dorset) 248–50
chromatography 78, 81, 84
Churchill, Winston 59, 241, 256
Civil Rights movement 99
Clarke, Bruce 269
class 21
climate 5, 6, 128–43, 144
 and IPCC 188
 and Lawson 241–7, 251–3

Index

and ocean ecosystems 221
The Revenge of Gaia 235–9
A Rough Ride to the Future 250–1
and Shell 106–11, 116–17
and Thatcher 229–31
and Tickell 227
Climategate 245
CO2 *see* carbon dioxide
CoEvolution Quarterly (magazine) 178, 209
Cold War 5, 61, 69, 217
Collins, Chris 228
Common Cold Research Unit (CCRU) 62–5, 67–9
Commoner, Barry 187, 188–9, 190–2, 193, 203, 210
Commonwealth Fund 222
communism 192
conscientious objection 49, 53, 58, 199
conservation 107
Coombe Mill (Dorset) 197–8, 225–6, 237, 239
Copenhage Climate Conference 244–5
Cornforth, John 72, 139–40
coronary disease 73
counter-terrorism 178
Critical Point, The (TV drama) 74, 182
Crutzen, Paul 2, 174, 193, 188, 189
cryobiology 74
cycling 163

Daisyworld 208–9, 221, 259, 269
Dalai Lama 217
Dale, Sir Henry 47, 53
Dartington Hall (Devon) 202–3
Darwin, Charles 94, 151, 166; *see also* neo-Darwinism
Dawkins, Richard 2
The Extended Phenotype 206–7
DDT 116, 121
Declaration on Global Change 236
Deevey, Edward 102
Delahunty, Mary 49–52, 56
Delingpole, James 257
dieldrin 121, 122, 123
Digby, Revd A.C. 27–8
dimethyl sulphide 163, 164, 167, 178, 221, 261
disequilibrium 92
Doolittle, Ford 4, 206, 265, 269
Dow Chemicals 173, 197
drugs 120–1
Dumbell, Keith 68
DuPont 169, 172–3, 184
Dutreuil, Sébastien 175, 269
dyscalculia 9, 70

Earth *see* Gaia theory
Earth Summit (1992) 231
Earth System Science 4, 233, 236–7

ECD (electron capture detector) 3, 5, 74, 75–7, 78
  and climate 137
  and Ireland 161, 162
  and secret service work 118–19, 196, 199
  and Vietnam 114–15
ecology 41
Ehrlich, Paul 2, 192
Ekholm, Nils Gustaf 128
electromagnetic theory 52
Elizabeth II of Great Britain, Queen 231, 237
endosymbiosis 151–2
Energy Development Corporation 217
Enlightenment 1, 44
Environmental Protection Agency 76
Epton, Sidney 119–20, 178–80, 181, 269
Ernest Shackleton, RRS 167, 168
espionage 2, 115, 117, 196, 260, 267
  and CFCs 162
  and Commoner 190–1
  and Halperin 192
  and Rothschild 110, 119, 145
  and tracing 162–3
eugenics 207
Evans, David 56
evolution 153, 206

exobiology 89, 96–9, 102
extraterrestrial life 82–3, 89, 90, 118; *see also* exobiology

Fairchild 81
Faraday, Michael 37, 175
fascism 41
FBI 190–1, 192
Fellgett, Peter 93, 94, 196
Finchley (London) 70–1, 72
fire experiments 58–9
First World War 19, 20–1, 27, 36, 59
Flannery, Tim 257
fluorocarbons 165, 174, 196
Foote, Eunice 128
fossil fuels 107, 108–9, 110, 128–36, 141–2, 242; *see also* gas industry; oil industry
Fourier, Jean-Baptiste Joseph 128
Frazer, Lorna 182–3
freons 169, 172–3

Gagarin, Yuri 82
GAGE (Global Atmospheric Gases Experiment) 161, 175, 212
Gaia theory 1, 2, 3–4, 6, 269–70
  The Ages of Gaia 221–2
  and beginnings 11, 17–18, 149–51
  and branding 205–6
  and Crutzen 189–90

Index

and Earth's atmosphere 93–4
and ECD 76
and Epton 179–80
Gaia: A New Look at Life on Earth 150–1, 182–6
and Global Forum 218
and Hitchcock 102–5
and Kirchner 210–12
and Latour 254–5
and Margulis 154, 178
and Midgley 239–40
and name 155–7, 158
and *New Scientist* article 180–2
and research 163–4
and scientists 206–8
and statue 213–15
and unveiling 158–60
and USA 209–10
Galileo Galilei 3
gas industry 108, 128, 236, 237, 242, 264
geo-engineering 6, 142, 244
geophysiology 232
Gibbons, Stella: *Cold Comfort Farm* 39
Giffin, C.E. 104
Global Forum of Spiritual and Parliamentary Leaders on Human Survival 217–18
Global Warming Policy Foundation 244, 245–6, 250–3

global warming *see* climate
Golding, William 2, 155, 156–8, 167–8
Goldsmith, Edward 226
Gorbachev, Mikhail 217
Gould, Stephen Jay 207
Great Depression 33
Greenaway, Dave 194–5
greenhouse effect 128–31, 140
Gribbin, John and Mary 258
Guerrero, Ricardo 220–1, 258
Guy's Hospital (London) 54

Haldane, J.B.S. 27, 171
Halperin, Israel 192
Hamilton, William 2, 207, 221
Hamilton Standard 91, 103
Harding, Stephan 226–7, 257
Harington, Charles 72, 112
Harvard Hospital *see* Common Cold Research Unit (CCRU)
Havel, Václav 232–3
Hawkes, Nigel 173
Hawking, Frank 59
Hawking, Stephen 59, 178
Heath, Edward 145
Hercules test flight 171–3
Hewlett-Packard 12, 81, 174–5, 208
Hickman, Leo 249
Hiroshima 60

Hitchcock, Dian 6, 80, 84–8, 89–91, 92–3, 94–101
  and Gaia theory 102–5, 151, 269
Hitchens, Christopher 2
Hitler, Adolf 48, 51
Hobsbawm, Eric 28
Hoover, J. Edgar 192
Horowitz, Norman 84, 94, 102
Houghton, Sir John 229
Howard, Frank 124
Hughes, Howard 86
Humboldt, Alexander von 166
Hutchinson, Evelyn 99
Huxley, Aldous: *Brave New World* 39
hydrogen 84
Hyslop, Helen *see* Lovelock, Helen (1st wife)

Icarus (journal) 96, 98
ICI 173
improvised explosive devices (IEDs) 195
Industrial Revolution 1
industry 42, 44–5, 165–6, 175–6
  and Gaia theory 183, 184
  and Nobel Symposium 187–193
intelligence services *see* secret service
interconnectedness 1–2
Intergovernmental Panel on Climage Change (IPCC) 188, 229, 243, 244

iodide 161, 163, 164
ionising detectors 75
IRA (Irish Republican Army) 5, 194–5, 200
Iran 195
Iraq 241, 248
Irish Troubles *see* Northern Ireland

Jagger, Mick 120
James Bond films 117
James, Tony 74
Japan 60, 233
Jenkins, David 142
'Jerusalem' (Blake) 36
Jet Propulsion Laboratory (JPL) 80–2, 83–5, 87–101, 174
Jones, Phil 245
Joseph, Lawrence E. 210–11

Kaplan, Lewis 93
Kennedy, John 82
KGB 67, 145, 190
Kirchner, James 210–11
Kloster, Knut 232
Kraft 73
Kumar, Satish 226, 257
Kump, Lee 259
Kyoto Protocol 242

Lane, John 202
Latour, Bruno 3, 158, 253–5, 269
  Moving Earths 259–60

Index

Lawson, Nigel 229, 241–6, 251–3, 257
leaded petrol 123–8
Lederberg, Joshua 81, 94, 102
Lenton, Tim 199, 259, 269
Letchworth Garden City 21–2
Lidwell, Owen 54
Lilly, John 86
Lipsky, Sandy 114
Liss, Peter 245, 259
livestock breeding 73–4
Lodge, James 141, 168
Lovelock, Andrew (son) 65, 71, 197, 208
  and drugs 120, 121
  and family life 213, 214
Lovelock, Christine (daughter) 58, 62–3, 65, 162, 197–8
  and drugs 120, 121
  and family life 70, 72, 213, 214
  and Lorna Frazer 182–3
  and maternal relationship 223–4
  and Sandy 220
  and wind farms 239
Lovelock, Helen (1st wife) 16, 17, 55–8, 198, 222–4
  and family life 213, 214
  and Finchley life 70–1, 72
  and Hitchcock 95–6, 99–100
  and multiple sclerosis 101, 201
Lovelock, James 1–10, 264–9, 270–1
  and accolades 231–3
  and bombmaking 31–3
  and CCRU 62–4, 68–70
  and centenary celebrations 255–60
  and CFCs 172–4
  and Chesil Beach 248–50
  and childhood 21–6, 28–30
  and climate 128–43, 144, 235–9, 250–3
  and Coombe Mill 225–6
  and Daisyworld 208–9
  and death workshops 204–5
  and drugs 120–1
  and early writings 41–5
  and ECD 75–7, 114–15
  and Epton 179–80
  and family life 71, 72
  and fatherhood 65–6
  and funding 174–6
  and Golding 156–8
  and Helen 55–8
  and higher education 52–3
  and Hitchcock assistance 103–4
  and ill health 176–8, 203–4, 262–4
  and Ireland 160–4

305

and JPL 80–2, 83–5, 87–8, 89–101
and Latour 253–4
and Lawson 242, 243–4, 245–7
and lectures 226–9
and literature 26–7
and Margulis 153–4
and maternal relationship 11–12, 15–17, 78–9
and MRC 73–5
and NIMR 46, 47–8, 53–5
and Nobel Symposium 187–90, 192–4
and Northern Ireland 194–5
and paternal relationship 33–4, 35–7
and pesticides 121–3
and Powys-Lybbe 201–3
and Rothschild 145–6, 147–8
and Sandy 216–17, 218–21, 222–2, 224–5
and schooling 27–8, 37–9, 40–1
and Second World War 58–9, 60–1
and secret service 118–20, 195–7, 198–200, 247–8
and Shell 106–14, 116–18
and southern hemisphere 166–70
and stratosphere 171–2
and tetraethyl lead 124–8
and Thatcher 229–31
and tracing 66–7
and USA 77–8
and women 48–52
see also Gaia theory
Lovelock, James (works):
The Ages of Gaia 221–2
Gaia: A New Look at Life on Earth 150–1, 182–6
Gaia: The Practical Science of Planetary Medicine 232
The Great Extinction (with Allaby) 204
The Greening of Mars (with Allaby) 204
Homage to Gaia 233–4
Novacene (with Appleyard) 255
The Revenge of Gaia 235–9
A Rough Ride to the Future 250–1
The Vanishing Face of Gaia 244
Lovelock, Jane (daughter) 58, 62–3, 65–6, 78, 198
and Finchley life 70–1, 72
and marriage 213
and Sandy 224
Lovelock, John (son) 71–2, 201, 213
Lovelock, Nellie (mother) 10, 11–21, 29–30, 39–40
and art shop 22–4
and death 200–1

Index

and Helen 56
and Mary Delahunty 51–2
and motherhood 27–8, 33–4
and widowhood 78–9
and Wiltshire 66
Lovelock, Sandra (Sandy) (2nd wife) 6–7, 216–17, 218–21, 222–5, 232, 233–4, 247, 263
  and centenary celebrations 255–60
  and Chesil Beach 248–50
  and Lawson 243
Lovelock, Thomas (Tom) (father) 15, 18, 20–1, 29–30, 34–5, 39–40
  and art shop 22–4
  and death 78
  and fatherhood 33–4, 35–7
  and Wiltshire 66
Lowell, Percival 83

Maathai, Wangari 217
McCarthy, Ray 169, 170, 172–3, 269
McCullock, Warren 92–3
MacIntosh, Frank 55, 60
Manchester 48–51, 52–3
Mandela, Nelson 217
Manufacturing Chemists Association (MCA) 170, 172–4, 175
March, Alice Emily (grandmother) 18–19, 21

March, Annie (aunt) 19
March, Florence (Florrie) (aunt) 14, 19
March, Frank (uncle) 19
March, Kate (Kit) (aunt) 14, 19
Margulis, Lynn 2, 150, 151–6, 158–60, 166, 177–8
  and climate 236–7
  and death 258
  and *Gaia: A New Look at Life on Earth* (Lovelock) 183–6
  and Gaia conferences 232
  and Gaia theory 181, 207–8, 269
  and Northern Ireland 200
  and ozone layer 172
  and Sandy 220–1, 226
  and scientific approach 240–1
Mars 5, 101–2
  and JPL 81, 82–4, 87, 88–90, 91–2, 94, 97–8
Martin, Archer 2, 72, 74–5, 77
Mason, John 135–6
Maxwell, Robert 168
May, Robert 4, 207, 265
MCA *see* Manufacturing Chemists Association (MCA)
Medical Research Council (MRC) 47, 63, 67, 70, 72–5, 78

medicine 5
Meghreblian, Robert 92
Met Office 135–6, 137, 138, 161, 231
MI5 114, 115, 164–5
MI6 12, 114, 267–8
Midgley, Mary 3, 239–40, 269
Midgley, Thomas 123, 124
military 53, 69, 81
Mill Hill *see* Medical Research Council (MRC)
Milne, Alasdair 145
Ministry of Agriculture, Fisheries and Food 73–4
Ministry of Defence (MoD) 64, 68, 69, 116
  and bombs 198–200
  and Northern Ireland 194
  and seaweed 164–5
  see also secret service
Mitsubishi 233
Miyazawa, Kiichi 233
Molina, Mario 170–1, 172, 189
Montreal Protocol 217
Montreal Treaty 173
moon 81, 82
Morton, Oliver 258
Mosley, Oswald 41
Mountbatten, Louis 200
MRC *see* Medical Research Council (MRC)
Muggeridge, Malcolm 143–4

Nagasaki 60
NASA 12, 86–7, 101–2; *see also* Jet Propulsion Laboratory (JPL)
Nash, Thomas 64
National Center for Atmospheric Research (NCAR) 131, 134, 140–2
National Front 207
National Institute for Medical Research (NIMR) 46–8, 53–5
Natural Environment Research Council (NERC) 135, 167
Nature (journal) 17–18, 92, 98, 103, 118, 137
  and CFCs 171, 172
  and southern hemisphere expedition 168
Needleman, Herbert 124
neo-Darwinism 153, 185, 206, 207, 209
Neumann, John von 175
New Age 4, 181, 205, 210
New Scientist (magazine) 180–2, 209
Newell, Homer 89
Newton, Edward 40–1
Newton, Isaac 218
nitrogen 75, 161, 187–8
Nobel Symposium 187–90, 192–4
Northern Ireland 178, 194–5, 199, 200, 217

Index

Norton, W.W. 222
nuclear power 5, 6, 60, 67, 107
  and climate 238–9, 240

Official Secrets Act 53, 160
oil industry 107, 111, 146–7, 203, 236, 237, 264
  and sceptics 242, 244
Omedes, Anna 258
Orchard, David 216, 224–5
Orchard, Sandra *see* Lovelock, Sandra (Sandy) (2nd wife)
Orpington (Kent) 33, 36–7
Orwell, George 39, 241
O'Sullivan, Michael 162, 175, 212
Oxford University Press (OUP) 182
Oyama, Vance 88, 91, 94, 102
ozone hole 5, 165, 171, 172–3, 188

pacifism 48, 49–50, 58, 69
Patterson, Clair Cameron 124
Pearce, Fred 258
pesticides 121–3
petrochemicals 5, 75, 77, 121–2; *see also* Shell
Philip, HRH Prince Duke of Edinburgh 251–3
Pitt-Rivers, Rosalind 74
planetary life *see* Gaia theory

politics 39, 144–5, 165, 207; *see also* Lawson, Nigel; Thatcher, Margaret
pollution 107, 122, 161, 164, 183
  and southern hemisphere 166–70
  see also greenhouse effect
polonium 67
Popjak, George 74
population rates 147
Porritt, Jonathon 226, 231, 240, 257, 268
Porter, Rodney 72
Porterfield, Dr J. 68
Portland, Henry Bentinck, Earl of 226
Porton Down 64, 68
Postgate, John 188, 189, 193
Powell, Charles 228, 230
Powys-Lybbe, Jenny 201–3, 204

Quakers 27, 32, 58

racism 207
Radford, Tim 256, 258
radiation 66–7
Rapley, Chris 2, 238, 244, 259
Reagan, Ronald 207
Rees, Martin 237, 259
religion 217
respiratory disease 63, 64
Reynolds, Orr 86–7

Ridley, Matt 245, 250–3, 257
Roberts, Walter Orr 141
robotics 81
rocket technology 81
Romanticism 44
Rosswall, Thomas 188, 193
Rothschild, Victor, Baron 106–7, 108–14, 116–18, 143–8, 234, 269
   and climate 128–9, 130–1, 134–6, 137–40, 141–3
   and death 258
   and Epton 178, 179
   and pesticides 122
   and secret service 115, 118–20
   and tetraethyl lead 127–8
Rowland, Sherwood 170–1, 172, 173, 189
Royal Air Force 171
Royal Dutch Shell *see* Shell
Royal Navy 69
Royal Society 59, 178
Runcie, Robert, Archbishop of Canterbury 217
Russia *see* Soviet Union

Saddam Hussein 248
Sagan, Carl 2, 81, 88–9, 93, 217
   and Margulis marriage 152, 153
   and Mars 102, 103
Sargent, Margaret 201

Schneider, Stephen 2, 176, 209, 229
Schrödinger, Erwin 94
Schumacher College (Devon) 226–7
science fiction 204
Science Museum (London) 24–5, 237, 250
Scott, Marilyn 56
seaweed 163–5
Second World War 41, 54, 58–61
   and NIMR 46–8
   and pacifism 49–50
secret service 69, 118–20, 195–7, 247–8
   and bombmaking 249–50
   see also espionage; MI5; MI6
Shell 12, 75, 76, 111–14, 117–18
   and climate 106–11, 116–17, 128–9, 134–5, 137–9, 141
   and Epton 178–80, 182
   and funding 174
   and Gaia theory 151, 184
   and pesticides 122, 123
   and pollution 164
   and secret service 118–20
Silicon Valley (CA) 81
Silverstein, Abe 80–1
Simmonds, Peter 73, 91, 92, 95, 99
sky 163

Index

smallpox 68
Smith, Audrey 74
Smith, John Maynard 4, 221, 232
smog 28, 131, 136–7, 188
Snyder, Gary: 'Gaia' 205–6
Soddy, Frederick 39
sodium bromide ions 53
South Africa 217
Soviet Union 21, 64, 196, 217
  and Commoner 190–1, 192
  and space exploration 81–2
  see also KGB
space exploration 60, 81–2, 83–4; *see also* Mars
Spanish Civil War 39, 41
spies *see* espionage
Standard Oil 124
Steel, Elaine 258
Strand School (London) 27–8, 33, 37–9, 40–3
suffragette movement 19, 30, 36
sulphur 76, 131, 133, 139, 142
  and seaweed 163, 164
Sutton, Graham 135, 138, 139

Tellus (journal) 159, 189
tetraethyl lead 116, 123–8
Thatcher, Margaret 145, 207, 241, 227–31, 256–7
Theresa, Mother 217

Thomas, Gordon 87, 90
Thornton 111, 117–20, 123, 124–5, 127, 178
Tibetan Book of the Dead, The 204
Tickell, Crispin 227, 229, 231, 232, 256
Todd, Alexander 53
tracing 54, 66–7, 114–15; *see also* ECD (electron capture detector)
Trivers, Robert 207
turbidity 136, 138
Turing, Alan 175

United States of America (USA) 76–8, 112
  and Gaia theory 209–12
  and Vietnam 114–15
  see also NASA
Uruguay 167
US National Oceanic and Atmospheric Administration 225
USSR *see* Soviet Union

V-2 rockets 58, 60, 81
Van den Ende, Marinus 55
Vengeance, HMS 69, 75, 196
Venus 81
Vietnam War 99, 114–15
virology 63, 64
Vishniac, Wolf 88, 91
Voyager programme 83–4, 90, 99, 101

Waldegrave, William 243, 257
Watson, Andrew 3, 208–9, 232, 259, 269
Watson, Robert 229
weaponry 60; *see also* biological weapons; bombs
Webb, Joan 74
Westwood, Vivienne 237, 258
W.G. Pye 12, 75
Whole Earth Catalog (magazine) 178
Wijkman, Anders 193
Wilde, Oscar 34–5

Williams, George 207
wind farms 238, 239
Winfrith (Dorset) 247–8
Wolf Trap 88, 91, 101–2
Wollaston Medal 233
World Wildlife Fund (WWF) 251–2
Wright, Peter: *Spycatcher* 145

Yasuda, Tetsuro 258